Memory Traces: Recursive Engrams

Memory Traces: Recursive Engrams

Ida Pearce, M.D.

Front Cover Illustration of the Turbine Effects (see page __) by Ida Pearce, M.D.
Back Cover Illustration of Processes Evoking Visual Perceptions (see page __) by Ida Pearce, M.D.

ISBN: 0996488200
ISBN 13: 9780996488204
Library of Congress Control Number: 2015914079
Stalbridge Press, Irvine, CA

Table of Contents

Introduction by Ida Pearce, M.D.

After some four decades in the practice of clinical ophthalmology, and following my serendipitous entoptic observations back in 1962 of the **Rotating Spiral and Target Waves**, my focus since October 2000 has centered upon explicit perceptions of Recursive Semantic Engrams, also known as implicit memory-traces.

My earlier annotated visual phenomena of Entopic Spiral Waves (dating from 1962) are documented in my 1963 correspondence with Dr. Ackerman at the Mayo Clinic. The source of those dynamic visual phenomena was not apparent until some decades later, finally identified as the *in vivo* manifestations of the reaction-diffusion phenomenon of Belouzov and Zhabotinsky.

In 2004, physicist Vladimir Vanag responded that my report to him of *the in vivo inward-rotating spiral waves* "was significant, and might be the first *anti-spirals* yet noted in a biological system."

The anomalous *levo*-chirality of 20 percent of these entoptic spiral-entities is also of current interest, dextro-rotation generally prevails in biology.

The **Lexical Engrams,** which I observed in 2000 were found to be retained faithfully and word-for-word - *verbatim et literatim.*

I first encountered the parallel system of **Dynamic Engrams** in September 2001.

My earlier dated **Entoptic Findings** have included the following:

- **Ocular Microcirculation** (1962), as evident in stroboscopic studies.
- **Spiral Waves** (1962–2012) witnessed while awakening during hypersynaptic hypnopompic-arousal episodes.
- **A Purple-Entoptic phenomena**, maximal at an imposed 40 Hz as noted since in 1962. This manifestation offers a simple means to assess acute fluctuations in ocular neurometabolism and the consequent prompt photo-regulation of ocular-perfusion levels. (See Riva C.E., 2005)
- **Transient Adaptive States** have supported Hypnopompic Images including Recursive Semantic Engrams (i.e., the memory-traces).
- **The Recursive Lexical Engrams**, as first witnessed in October 2000, are retained intact, *verbatim et literatim.*
- **The Dynamic Spatial-Motion Engrams** I have witnessed entopically since September 2001.

Findings of these two visual systems' phenomena provide valuable insights into the innate, hardwired dynamics of neural systems and to the semantic, inculcate software, which then enables literacy. These recursive engrams are focally legible, *verbatim et literatim.*

This book has two main parts. Part One is Entopic Manifestations of the Hardwired Visual Systems, and Part Two is Entopic Manifestations of the Semantic Software.

Topics are labeled A through F, with subparts. I wrote Topic A first, based on my findings and observations and continued with Topics C through F.

In later writing this book, I have reorganized the subsections for continuity of subject but have kept the original labeling, which reflects the order of my discovery of the material.

Topic A1: "Ocular Pressure Phosphenes and the Microcirculation Dynamics"
Topic A2: "Visual Templates as Perceived in Vigilant States"
Topic A3: "Synesthesia and Synkinesia: Autonomous Vertical Ocular Saccades"
Topic B5: "Recursive Lexical Engrams"
Topic C6: "Meandering Targets Can Reveal Edge Detection"
Topic C7: "The Dynamic Engrams (Motion-Memory)"
Topic D8: "The Blind Spots of Mariotte Serve as Landmarks"
Topic D9: "Geometry—The Eight-Fold Way?"
Topic D10: "Mandala: The Subjective Polar-Radial Array"
Topic D11: "The Helmholtz Traveling Waves"
Topic D12: "Spiral Wave Dynamics in Neocortex, Explicit Perceptions *In Vivo*"
Topic E13: "Purple: A Nonspectral Color,"
Topic E14 "The Dominant Purple Manifestations vs. the Subjugate Yellow-Green Perceptions"
Topic E15: "The Prompt Ocular-Perfusion Marker Offers Applications in Glaucoma Management"
Topic E16: "Photic After-Images, Quasi Scotomas: "Ionic Waves and Pixels"
Topic F17: "Visual Perceptions of Vestibular Signals,"
Topic F18: "Ambiguity and Biases."

Key Terms and Concepts

"When I use a word ... it means just what I choose it to mean."

Humpty Dumpty to Alice in Lewis Carroll's
Through the Looking-Glass

"Once an imperative for all scholars, the underlying significance of common technical words is becoming blurred." Shelfer L.F., Isaacs J.T., 2008

Corruptions of words, euphemisms, and ambiguities are common in vernacular parlance.

Morphing of terms follows technical progress...and with some interdisciplinary divergences.

Biasing versus Priming

Instruction Given = Inculcation Imposed

Learning of a task: Mnemonic Rentention with Recall or Iterations
Priming: Generally and historically taken to mean a setup or preparation for a forthcoming or anticipated event. However, *priming* was defined by Tulving and Schacter, in 1990, as the

"implicit learning of highly specific data, secured by a Perceptual Representational System."

In the present context, the installation of a semantic engram is understood as an acquisition, not as a priming. However, a visual task is generally a continuous viewing of one congruent event within an epoch of time or the reading of a page of text. Tasks may be imposed by a research protocol, such as a restriction.

Adaptation: A function colloquially and generally understood as the making of systematic modifications in order to obtain a new condition or to enable a secondary purpose.

Barraza (2008): "If the purpose of adaptation is to fit sensory systems to different environments, it may implement an optimization of the system."

Photic Adaptation: The changes in gain or in the dimensions of the receptive fields.

Such adaptation is a process common in biological evolution and is seen also in technological evolution and in common speech. Terms are subjected to misapplications, for instance:

The use of the term *adaptation* in the Motion After-Effect context suggests that the MAE has a role other than that of simple resets of cortical motion functions; thus, the functions are comparable to the color-opponency resets of the common contoured after-images.

Instruction: Semantic input presented to sensorimotor systems; items arguably not necessarily proven retained.

Training: Inculcation of motor loops in response to imposed multimodal signals: *entrainment.*

Inculcation: Semantic input successfully preserved in the sensory system.

Learning: May be defined as retention of a semantic input; deemed accomplished when thereafter this specific item may be verified by recall, by autonomous engram iteration, or implicated by behavior modification or inferred from some cerebral activities as detected with electrotechnology.

Mnemonic Retention: The acquisition of a semantic engram is fundamentally a banking of information as a short- or a long-term deposit, rather than a "priming."

Recall: Successful, specific deliberative search.

Recursions: The autonomous reiteration of engrams.

Encoding: I understand this term may at times refer specifically to the automatic—and inevitable—acquisitions of the autonomous engrams. *(See Halpern, B.P., 2000.)*

Reading: Goal-oriented discriminations within scalar hierarchies.

R.H. Baud (2007): "In order to encourage appropriate usage of terminologies, guidelines are presented advocating the simultaneous publication of pragmatic vocabularies, supported by terminological material based on adequate ontological analysis."

R.G. Jahn and B.J. Dunne (2007) recommend "a more explicit and profound use of interdisciplinary metaphors."

Richard Semon, in 1904, provided terms of his own invention—*Engraphy*, *Engram*, and *Ecphory*—in order "to avoid the potentially misleading connotations of ordinary language."

In this engram context, Cursors, Scanners, and Lexors" refer to specific afferent sensors with noted angular projections of one, five, or ten degrees.

Part One

Entopic Manifestations of the Hardwired Visual Systems

Ocular Pressure Phosphenes and the Microcirculation Dynamics (Topic A1)

Visible Structures: Well-recognized in the literature, the entoptically visible ocular anatomical structures include precorneal films, lenticular and vitreal opacities, the shadows of the retinal vascular arcades, and the preretinal capillary net.

Entoptic Circulatory Events: Corpuscular capillary and arteriolar flow, the effects of hypoxia, can be demonstrated subjectively by steady or by intermittent illumination, by local or global pressure, and by systemic circulatory or postural manipulations.

Auto-regulations to briefly imposed ocular pressure: This is a fatiguable one-time-only response demonstrable by the recovery of the initial amaurosis within a ten- to fifteen-second delay and is again demonstrable only after several minutes respite.

Phosphenes: Generated by mechanical ocular stimuli; vigorous sustained massage.

The following illustration of the subsequent entoptic perceptions depicts

1) A bright yellow ellipse at the macula, represented here as a light grey ring in the center, clearly delineated.
2) A diffuse yellow glow or mosaic, represented in grey, pervades the general field, notably absent in the four oblique quadrants.

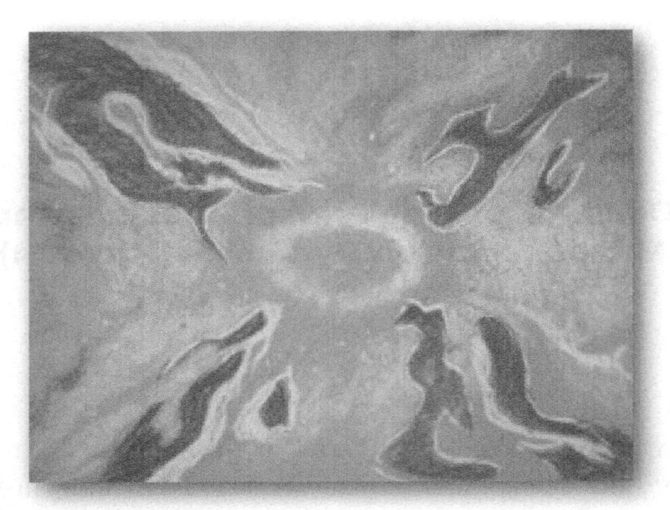

Stiles-Crawford Effect This topography suggests that distortion of the globe is minimized at the sites of the orthogonal rectus muscles (yellow areas) with sufficient retinal corrugation displacement or mild edema in the unsupported regions to incur a **Stiles-Crawford effect.**

The oval ring at the macula is evident. According to P.S. Berstein (2004), "Resonance Raman measurement of macular carotenoids in the living human eye."

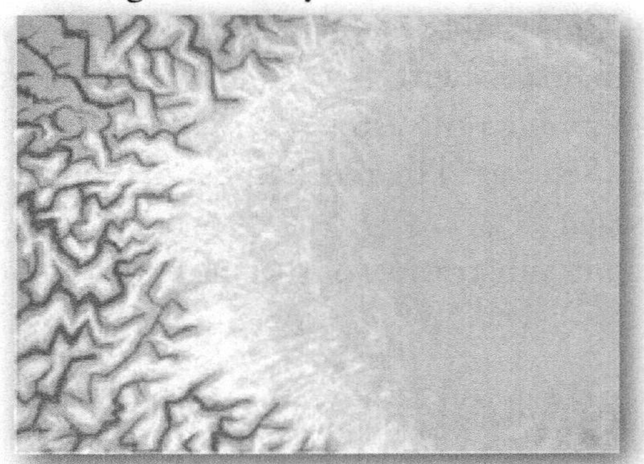

Helmholtz Traveling Waves Another effect was seen in recumbence with oblique shafts of setting unlighting, incurring a Stiles-Crawford effect with the incidental onset of Helmholtz Traveling Waves.

Visual Templates as Perceived in Vigilant States (Topic A2)

The gameboard or template upon which vision is inscribed (the Mandala platform) is tethered syntactically to this common postural and gravitational field. (The phenomenal morphology and kinetics of the Mandala are further described in "Mandala: The Subjective Polar-Radial Array.") Thus, online visual signals have instant confirmation of their image projections into the three-dimensional space by collaboration with the "instantaneous" (80 ms) signals from the vestibular and proprioceptive systems.

The vestibular signals outrun the laggard 120 ms retino-oculomotor responses. The cognate signals from all three sources are registered almost simultaneously, that is, within the temporal neural constraints of their specific sensory networks of conscious perception or default systems. (See "Visual Perceptions of Vestibular Signals" in Part Two.)

According to F. *Pulvermüller (2009)*, "The full activation or ignition of specifically distributed binding circuits explains the near-simultaneity of early neurophysiological indexes of lexical, syntactic and semantic processing."

Templates Perceived as Hypnagogic and Hypnopompic Images

While passing from sleep-to-waking-to-sleep "interludal" visual experiences are not rare; fragmented images are often witnessed when "*in limbo*."

Ohayon in 1996 found 20 percent of his five thousand telephone respondents claimed the hypnagogic images noted predominately in their wake-to-sleep transitions.

Stickgold in 2000 reports the hypnopompic recall memories of a dozen subjects, both normal and amnesiac. Few other studies have attempted to link the role of a brain system in sleep-to-waking transitions seeking a role in cognitive performance during waking status in limbo.

The rapid descents into sleep may hamper effective recalling of such fleeting visions, but the sleep-to-wake (hypnopompic) episodes allow prompt documentation by an awakening and motivated investigator. Engrams with stereotypic semantic lexical content are of major interest.

The various images as characterized here are distinguished by their specific patterns or contents, modes of onset and offset, longevities, their formal rendering styles and/or motilities, and semantic lexical content.

Instantaneous onsets or emergent images may assemble by aggregation of particles over about three seconds.

Domains. The cognitive field extents witnessed in limbo seldom extend beyond thirty to forty degrees.

Evolutions. Semantically unrelated images may compete for segments of the overall cognitive visual space; usually there is switching of complete images or insertions of map fragments, beginning at the focal center.

Scannability. Illusory apparitions described by Gillespie in 1989 as "scannable" were of a "two-dimensional flatness, nonrepresentational and unrelated to other images" (Pinker S., 1998).

Mental image scanning is a process distinct from eye movements or eye-movement commands. A perception of scannability may be a function of cognitive attentional direction.

Audio-visual Synesthesia. Online music or intrusive sudden sounds may precipitate novel images or change the dimensions, vigor, or content of current illusory images. In limbo, audio drives visual.

Fluid waves, vortices and streams, or currents, spiral and traveling waves, are seen arising from chaotic particulate motion.

Textural and Turing Patterns. With allover patterns of repetitive idiomorphs on Cartesian or polar coordinates, each idiomorph retains its own domain while displaying synchronized motility.

Pictorial Representations, Landscapes or Objects, Faces or Figures. The grotesque facial distortions typical of the Charles Bonnet syndrome suggest that fragments of images arise from the cortical facial-feature file recognition center.

Manifestations of the Recursive Lexical Engrams are recency and load-dependent. (See "Recursive Lexical Engrams.")

Also seen "while awakening in limbo" are the rotating spiral waves and their transformations, viewed during the synaptic-hyperactive generation of reaction-diffusion systems, and also an occasional entoptic Turing pattern. (See "Spiral Waves.")

Floral Fibonacchi

A geometric drawing representing a spider web and the "the circus ternus."

According to I.R. Epstein (2009), "Amplitude modulation of spirals can evolve to chambered spirals resembling those found in nature, such as pine cones and sunflowers." And spider webs and shells?

Meandering Targets Can Reveal Edge Detection (Topic C6)

A Circus Ternus Target: In action this demonstrates Arnold Tongues, Chiral Motion, Contextual Associative Fields, and the Kanizsa Anisotropic Effect.

Design: This target design comprises ten concentric rings with sixteen sectors of alternating black and white tiles within a 14 cm diameter circle to be viewed at a distance of one-half meter.

"Circus Ternus" targets are viewed binocularly with hand-held oscillatory motions imposed upon these targets. Such stimuli readily elicit Arnol'd Tongue Percepts and Illusory Rotary Motions.

The angular subtenses of these areas (tiles or elements) are believed to be proportional to the architectural array of the visual template. The dartboard displays are presented to the observer en face, at a distance of one meter, and are thereafter observed with changing ambient luminances and aspect-ratios, and in combinations of rotational frequency when tilted at forty-five degrees.

See "The Polar Mandala," Topic D10, represented in the planforms published by Bressloff in 2002 described by Purkinje as the Rosettfigur. Bell, Dickinson, and Badcock, in 2008, note "Radial frequency adaptation," which suggests polar-based coding of local shape cues.

Linkage of Spiral Wave Dynamics in Neocortex and a Motile Target

This Circus Ternus Target was designed to further explore the cerebral motion detectors, the neural systems which purportedly had enabled my subjective perceptions in 1967 of the endogenous rotary spiral dynamics.

Perceptions of Rotary Motion. Seiffert, in 1998, evaluating perceptions of the exogenous "minimal displacement of the pattern of annular rotating gratings oscillating at various rates," concluded that "the changing position of features was more readily detected than was velocity."

Visual perceptions are attained online, and also off-line as perceptions of intrinsic images.

Biological spiral wave activities, as identified in simple organisms and in vitro neural tissues, arise also in brain tissue.

These activities may be detected subjectively as Rotating Spiral Waves (RSW). These in vivo entoptic phenomena, which I had chanced upon in 1962, were later identified with the Belousov and Zhabotinsky reaction-diffusion systems.

The RSW are seen entoptically, with multiple discrete cells rotating at about 10 Hz for up to four seconds. Some cells reach diameters up to three- to four-degree angular subtense. Since they are narrow, alternating darker and lighter bands of a cell, are feature-less, and putatively are equidistant, how then can their rotations be distinguished and enumerated?

This dilemma was addressed by proposing the entopic RSW phenomenal motions as visually equivalent to those of a spinning top or hula-hoop dancer, as viewed while rotating, with the precessions only distinguishable with a meandering of its pivotal point.

This concentric geometric array is mounted on a handheld fan and manipulated with precessing sinusoidal oscillations between 1–8 Hz, obtaining an approximately forty-five-degree tilt and tip of the target. When this target is presented in certain motions, arrays of Arnol'd Tongues and of Turbine Effects are generated. These resonant effects are absent when using a mechanical turntable with the spindle of the machine aligned with the axial center of the target.

Arnold Tongues, Reciprocal Antagonism. According to Serrano-Pedraza (2007), "Resonance Evidence of Reciprocal Antagonism between motion sensors tuned to coarse and fine features."

Arnold's Tongues are seen at the lower frequencies; the "Turbines" appear with acceleration. Both effects have increased salience at low luminances and extended domains with accelerations. Only a meandering, off-axis rotation of the Circus target, elicits these resonant Arnol'd Tongues, or the Antagonistic Chiral Motions.

Detection of rotation of the entoptic RSW is deemed to be enabled by a slight shifting eccentricity. RSW are known to meander either inward or outwardly; reportedly the ensuing chirality is contrary to that of the meander.

Only a meandering off-axis rotation presentation of the Circus target elicits the perceptions of Resonant Arnol'd Tongues of Antagonistic Chiral Motions and of Color Asynchrony. These effects had been unanticipated.

According to Bressloff, in 1999, "Based on anatomical evidence, we assume that the lateral connectivity between hypercolumns exhibits symmetries, rendering it invariant under the action of the Euclidean group E(2), composed of reflections and translations in the plane, and a (novel) shift-twist action."

J.E. Dickinson and D.R. Badcock, in 2007, found that "Selectivity for coherence in radial and concentric, +45 degrees and -45 degrees spiral/polar orientation in human form vision."

Present Findings

Complex hybrid motile patterns appeared in different zones of the Circus target, most apparent in its elliptic motions and with oscillations between 1–8 Hz, together with a forty-five-degree tilt and tip of the handheld fan. These transformations depend upon velocity, luminance, and adaptation. Patterned motions arise from oscillation or linear planar dynamics, and their intersections generate resonances and spiral convolutions.

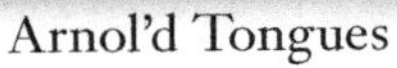

Arnol'd Tongues

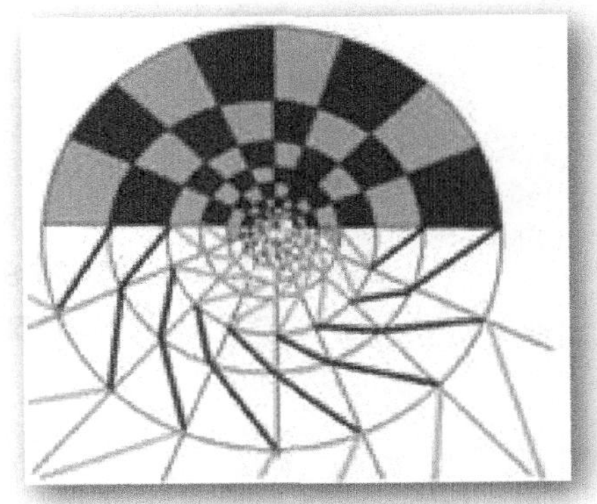

Turbine Effects

Meese T.S. and Anderson S.J., 2002, conclude that the detection of complex motion in human vision requires both cardinal and spiral mechanisms with a half-bandwidth of approximately forty-six degrees.

Turbine Effects (Serrano-Pedraza, 2007) present "evidence for Reciprocal Antagonism between motion sensors tuned to coarse and to fine features."

Confirmed in experiments with orientation-filtered noise. If the 1 cycle/deg noise flickered and the 3 cycles/deg noise moved, the 1 cycle/deg noise appeared to move in the opposite direction to the 3 cycles/deg noise even at long durations. These transformations depend upon velocity, luminance, and adaptation.

The Blind Spots of Mariotte Serve as Landmarks (Topic D8)

The locations of the optic nerve heads can be identified as asymmetric positive/negative entoptic phenomena. The imple-tion mechanisms of these "scotomata" (as described by Pearce, I., *Arch. Ophthal.* Vol. 79, May 1968) relate to cortical functions; similar mechanisms also erase the momentary perceptions of the preretinal vascular shadows.

These intraocular vascular retinotopic images are "object-derived"; hence, they are cognitively transparent to the exogenic, homotopic images, as both images are transmitted along the same online pathways. But the complex illusory images, which are generated by neural processes in the higher visual pathways and derived from "percepts in storage," engrams, dreams, and the hypnagogic images, are generally not transparent with external images. The subjective recognition of entopic structures is useful in the evaluation and scaling of other less familiar entoptic visual phenomena.

With both eyes closed but lit (an umbral view), the locations of the optic nerve heads can be identified by manually occluding alternate eyes, thus manifesting asymmetric positive entoptic phenomena. The impletion mechanisms of these "scotomata" (as described by Pearce I., *Arch. Ophthal.*, 1968) relate to cortical functions; similar mechanisms also erase the momentary perceptions of the preretinal vascular shadows. These images can act as landmarks and as scalar references for other entopic phenomena.

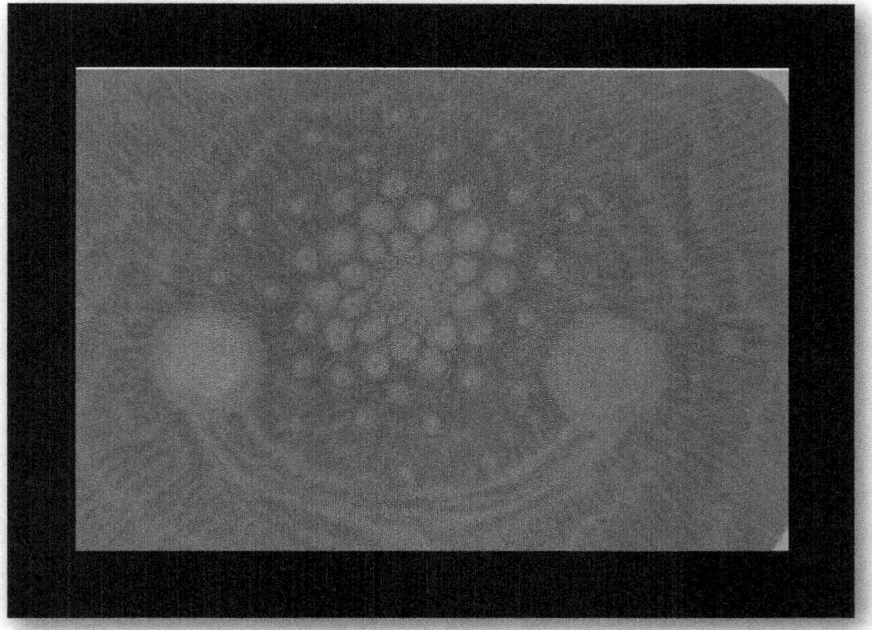

Entoptic Landmarks

The intraocular retinotopic images are "object-derived" and hence are cognitively transparent to the exogenic, homotopic images, as both images are transmitted along the same online pathways. But the complex illusory images, which are generated by neural processes in the higher visual pathways and derived from "percepts in storage," engrams, dreams, and hypnagogic images, are generally not transparent with external images.

The Geometry of Migraine Auras (Topic D9)

With Observations of a Counter-Flow Dynamic, 1990

Descriptions of the scintillating spectra date back to Roman times. In 2002, Bressloff and Cowan review the micro-anatomical dimensions and topology of the visual cortex. Ohki, in 2000, describes "rows of clockwise and counter-clockwise pinwheel centers arranged alternately in a unique geometrical pattern." The migraine aura has features commending it as "a cellular model of excitable media including curvature and dispersion," as defined by Gerhardt M., 1990. Migraine phosphenes and the retino-cortical magnification factor are defined by Grusser O.J., 1995.

Subjectively, the apical angles of the migraine aura-chevrons appear to widen as the process spreads peripherally, suggesting the anisotropic domains of the neural template (see "Meandering Targets Can Reveal Edge Detection," Topic C6) consistent with the known dimensions of receptive fields—a Fibonacci series? The domains so depicted appear broader circumferentially than in their radial measure, giving a flattening effect on the apical angles of a progressively centrifugal chevron display. No diminution in their number is evident, though jittering is increasingly evident as the magnification progresses.

The uniform morphology of each of the visible chevron elements, or textons, is consistent with the known parameters of visual cortical neuroarchitecture.

Further Considerations of the Visual Phenomena:

1) Triggers, neurovascular, undetermined.
2) Vascular components.
3) Spectral range.
4) Subjective morphology: Adjacent chevrons? Baseless triangles? Conjoined zigzags?
5) Anatomical dimensions and topology; rate of cortical spreading approximately 15 cm in twenty minutes.
6) Counter-rotating micro-elements, as seen March 14, 2001, at 1:30 p.m.
7) Why macular sparing? Why never a foveolar onset? It is remarkable that the much-emphasized spreading depression that follows activation is not initiated at fixation.
8) Why no callosal transmissions (V1/V2 borders cross the midline, Bressloff, 2002).
9) At eccentricity fifteen degrees, there is no evidence of a blind-spot absence of the aura.

Mandala: The Subjective Polar-Radial Array (Topic D10)

A Global Percept: The "Mandala" Template—The Purkinje Rose

The term "Mandala" is a tag used to identify the subjectively perceived rose-window geometry. After adaptation of both eyes to bright sunlight, a subjective pattern soon appears in dark; it is seen as a radial figure of motile luminous blue nodes on a darker ground and persists indefinitely and with modulations.

Adaptive Transients: Empirically demonstrable by visually cycling between red umbral light (Fig. 1) and darkness (Fig. 2) at seven-second intervals; adaptive transients generate red/blue offset radial patterns whose salience increases to a maximum during each of the seven-second periods, and this effect is further enhanced by continued cycling.

After exposure to daylight, a subjective pattern of oscillating clusters is commonly experienced in the dark. This Mandala, the master template of radial pattern, reverberates throughout the visual hierarchy. The cortical network oscillation is organized with two settings tightly coupled to form locally opponent oscillators. (See E13 "Purple: A Nonspectral Color," E15 "The Prompt Ocular-Perfusion Marker Offers

Applications in Glaucoma Management," and D11 "The Helmholtz Traveling Waves.")

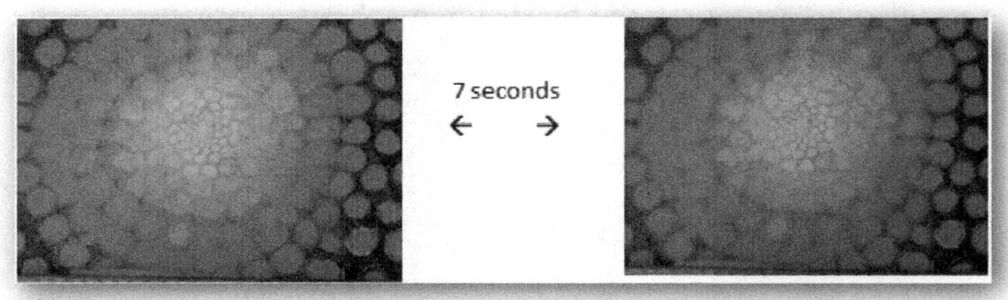

Entoptic view, with closed, lit lids, is red in color.

Entoptic view, in full occlusion, is blue in color.

Entoptically, the dynamical visual system appears as a polar network of pulse-couple oscillators in a radial lattice, in which unstable attractors can arise from dichoptic, asymmetric photic stimuli; synchronized oscillators (clusters) are activated, resulting in apparent interocular transfers or combinations. These Mandala oscillatory phenomena can be modified by changes in lighting, local or global, symmetric or dichoptic.

Dark adapted, the Mandala pattern is unscannable and appears as radially symmetrical luminous nodes centered upon the foveolar fixation point, a polar geometry familiar to most cultures and depicted in the ancient Mandalas, serving also as the nautical compass.

The streams of signals emitted by these "luminous" nodes, estimated at between 5–10 Hz, are experienced in the dark as "the visual noise" of syncrhonized oscillating clusters. The local intensities are modifiable by cyclical sweeps of the Helmholtz Bands (see D11 "The Helmholtz Traveling

Waves") by local perturbations and by asymmetric diffuse ambient light.

Pressure Effects. The polar topography in the umbral view may also be demonstrated by steady mechanical pressure on the eye:

Entoptic view with closed, lit lids. Schematic Mandala

Geometry—The Eight-Fold Way?

The Jebwa tribe constructed dream catchers by tying sinew strands in a web around a small round or tear-shaped frame of willow, similar to their technique for making snowshoe webbing. Their designs echo the Mandala retino-cortical template, commonly with eight or six vertices.

Flicker with a strobe lamp. Toggling briskly between 20–40 f.p.s., some dark and light standing waves appeared with wavelengths then estimated at 30–300 microns.

The Helmholtz Traveling Waves – HTW (Topic D11)

These Global Waves Traverse the Visual Field in Four Seconds

Eldridge Green, in his 1920 paper, described "currents seen in the dark…bluish violet circles in the periphery…which…advance to the center of vision, break and brighten at the fovea…and are succeeded by a second circle." These phenomena were named earlier by Helmholtz as blue bands, which he noted to have a rhythmic activity linked to respiratory cycles.

Ermentrout and Kleinfeld, in 2001, remarked that traveling electrical waves in (excised) cortex "are typically present during periods outside of stimulation." This is presumably applicable to exclusion of any environmental inputs that might engender frequency variances.

Steven Strogatz, in 2002, called to attention Art Winfree's original concept of "gigantic systems of weakly-coupled oscillators with randomly distributed frequencies…As the variance of frequencies is reduced the oscillators remain incoherent until a critical threshold is reached when the oscillators synchronize spontaneously."

An oscillatory synchronization may be witnessed subjectively in the Helmholtz Blue Band effect, a visual phenomenon wherein bichromic traveling bands are seen traversing the cognitive field as circular, elliptical, or hyperbolic contours Helmholtz Traveling Waves: HTW.

Induction Methods of HTW and Personal Observations from 1962

Induction: The Helmholtz Blue Bands (global traveling waves). These are most salient following activity outdoors in sunshine; then in a dark-adaptive process these waves emerge to view for the now sedentary observer. After adaptation to bright ambient light, retinal stimulation by diffuse or focal light (or by local pressure on a globe), an observer who then remains at rest with eyes closed in the dark may shortly witness the Helmholtz Traveling Waves, HTW.

These luminous annular bands arise within a minute and can iterate for about two minutes; each band transits the visual field within four seconds. These bands pass across the full cognitive field, appearing as pairs of bichromic luminous zones; only two such opponent pairs are present at any one time. Each blue-purple zone is paired with a dimmer green-yellow zone of equal width. Each set is refreshed at approximately two-second intervals, and together they travel continuously at a steady pace, either centrifugally or centripetally, and with either circular, elliptical, or hyperbolic contours transiting the field within four seconds, this configuration may be determined by retinal stimulation by diffuse or focal light or by local pressure on a globe.

March 31, 2007. When arousing from sleep while dark adapted, the HTW arise from increasing synaptic activity of arousal, but the centripetal bands are then of diminished luminance and of brief duration.

Propagation Speed: As if mapped onto a retinal surface, the apparent propagation speed of the retinal Helmholtz "traveling waves" is 5–10 mm per second. There is no perception of the acceleration as seen in evolution of the migraine aura, attributable to retino-cortical magnification. Blake, in 2001, writing on the dynamics of one-dimensional traveling waves in visual perception, notes that "when mapped onto visual cortex, propagation speed was computed at 24 mm/sec across VI" and that the propagation speed of such traveling waves is independent of eccentricity.

Cycling Rate: The slow rate of HTW intra-retinal cycling matches that of the automatic regulated respiration rate, and is comparable to the periodicity of visual retinal rivalry seen in reversals of ambiguous images, also commonly around four seconds.

Hue and Kinetics. A purple and yellow-green bistability is reported elsewhere (Topics E13 and E14) in illusory phenomena. Welpe, in 1978, and Schurcliff, in 1959, identified it as sourced in the retina. The luminous purple-blue segment of the bichromic HTW band is usually untextured, waning in intensity from leading to trailing edge. At the rear it gives way to the darker reticulated and flickering yellow-green zone; this green zone is then overtaken by the next oncoming blue-purple wave front.

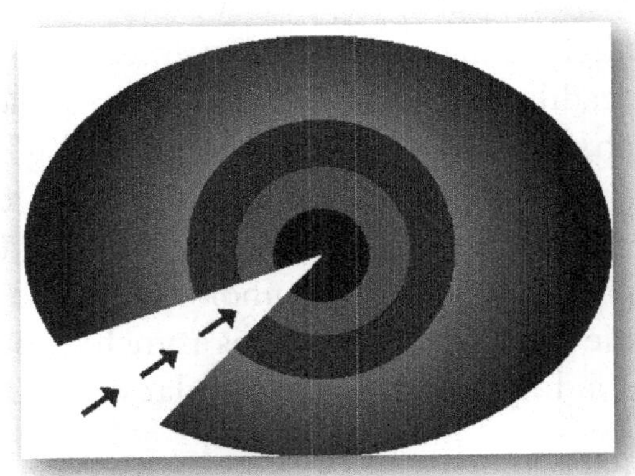

Convergent centripetal concentric HTW Blue Bands, represented here in grey.

Each HTW front transits "from source to sink" in about three to four seconds. Each sinusoidal band is generated anew at rates varying between one and three seconds and disappears in a "sink" located in the visual field diametrically opposite its apparent source of its origin.

These luminous bands arise from the "hot spots" selected for priming upon either retina. With a panoramic light source, the ensuing annular bands then converge upon the fovea where they disappear, as was described by Green in 1920. However, following a focal macular irradiation, the bands are circular, expanding and radiating away from the foveally imprinted "source." If the stimulus was restricted to one half or one quadrant of a monocular field, the expanding contours then appear in parallel as elliptical, parabolic, or arcuate contours as they then traverse the full visual field by imposing asymmetric conditions and bi-retinal neuroconvergence can be demonstrated.

1) With patching one eye (the right selected) while the left open eye is engaged upon a normal reading activity, the patched right eye shortly perceives diffuse fine particulate luminous agitation, an "interocular transfer" (Mackay). This oscillatory activity is most salient when a low level of diffuse light is then permitted to enter the "occluded" eye. Now steadily fixing the left gaze upon some other target or graphic, waves of "perceptual blanking" may traverse the cognitive field. On then closing both eyes, this wavering phenomenon may reappear with the typical tempo and topography of the Helmholtz waves.
2) By photic priming of the retinas asynchronously or at non-corresponding points (focal macular versus any peripheral

site) or by applying local pressure to only one globe, the consequently mismatched topologies and phases of the contralateral displays can then easily identify each retinal contribution to the perceptually discrepant displays. When thus induced asymmetrically, both the choreographies and the intensities of their respective bands can be distinguished, either by their being mutually out-of-phase and/or topologically distinct.

Yet after a few cycles, these competing patterns no longer resemble the simply overlaid conical sectors (one, the circular, the second, the hyperbolic pattern as described above). Shortly the wave fronts combine to follow an apparently spiraling vector so that the display now resembles a scimitar-shaped sector that rotates around the central point and may so continue for several cycles.

Wilson HR, Kim J.: "Perceived motion in the vector sum direction...the visual system combines...components to produce perceived motion in the vector sum direction, even when this deviates by as much as 53 deg from the intersection of constraints direction."

Modulations of Other Phenomena by the HTW

3) The HTW bands contain opponent hues, these zones of equal widths, and these modulate the Mandala rosette pattern, the fixed radial array over which these HTW "currents" flow (see The "Mandala" Template—The Purkinje Rose, aka Bresloff's planform).

The blue band in the more luminous phase appears to further brighten and intensify the "already bright nodes" of the Rosette pattern. But the succeeding yellow-green phase, the less luminous phase of the HTW, rather than appearing to darken, the brighter figures of the array brighten only the "darker background" of an already established overall spatially stable Mandala pattern. Thus, in addition to a hue-change effect across the visual field, the luminous blue band enhances the pattern's contrast, whereas the yellow-green activity diminishes the perceived contrast, and together the two phases may create a three-dimensional light shadow aspect to the illusory figures.

Smythies J., 1958–2000: "This sculptural effect is seen with closure of both eyes and is sustainable for minutes."

Shevelev I.A., 2000: The scanning of the visual cortex by a spreading wave process operating at the frequency of the alpha-rhythm reads information from the visual cortex.

Modulations of the HTW by Other Phenomena

4) HTW modulated by respiration-brainstem activity, the system of bands transiting from source to sink over three to four seconds, are synchronized with the respiratory rhythm rate, as was noted by Helmholtz. However, when respiration is voluntarily suspended, the phasic sweep rhythm continues unchanged for a few cycles before the HTW cease. This indicates that central neural rather than the peripheral mechanical effects of respiration modulate the perturbation that is imposed locally or monocularly by the photo-adaptive triggers.

Grossman P., 2007: Respiratory sinus arrhythmia plays a primary role in regulation of energy exchange by means of synchronizing respiratory and cardiovascular processes during metabolic and behavioral change.

McGuinness M., 2004: Arnold tongues in human cardiorespiratory systems. While RSA can cause synchronization where inspiration modulates heart rate, the strongest mechanism for synchronization is CVC, cardioventilatory coupling (CVC, where the heart is a pacemaker for respiration).

5) HTW modulated by vestibular inputs. Smooth pursuit of a selected uprising arcuate wave front can be carried out—or simulated—by smooth elevation of the chin over a physically convenient 120-degree angle. This pursuit succeeds in tracking, "stabilizing" a selected wave front for some five seconds; thereafter, the perceived motions of the oncoming wave fronts resume across the cognitive field, again traveling upward relative to borders of the cognitive field. (A similar phenomenon has been observed with smooth pursuit of illusory persistent motion.)

Nicola E.M., 2002: Drifting pattern domains, DPDs, exist due to a locking of the interface velocities, which is imposed by the absence of space-time defects near these interfaces.

6) Oculomotion tends to disrupt HTW. Disruption of retinotopically based TW by oculomotion was noted also by Blake.

What Neuro-Retinal Mechanisms May Enable Perception of the HTW?

It appears that concentric waves of alternating polarity drive the mechanism initiating the perturbations of the neurovisual system. The size of these cyclical concentric zones (some forty-five-degree subtense) would preclude their attribution to identifiable neuronal pools or networks in the retina itself, for these are not of comparable extent.

In the dark-adapted eye, the pigment epithelium participates in the electric potential differences in ERG measurements that waves of sequential firings of opponent elements in a second order network might coincide with, or cause, or result from, some transient electro-motive potential changes across this p.e. "membrane." However, the cycling rate of the HTW is much slower, so it is problematic to attribute the phenomenon to this purely local mechanism, or to account for the HTW bichromaticity. (See E15 "The Prompt Ocular-Perfusion Marker Offers Applications in Glaucoma Management"concerning bistable opponencies and flux.)

However, depolarization of the photoreceptors in the dark releases a neurotransmitter to horizontal cells and to bipolar cells. The ganglion cells (Dacey, 1996) do exhibit spectral opponency, but it is said that the horizontal cells. H1 obtain input from L and M cones only, and that H2 input is mainly from S cones and that no opponent transformation was identified in either H cell type. "Horizontal cells of the primate retina: cone specificity without spectral opponency." Are we seeing differential firing sequences arising from these two populations of horizontal cells? (See E15 "The Prompt Ocular-Perfusion Marker Offers Applications in Glaucoma Management" concerning bistable opponencies and flux.)

Conclusion

Based on the geometry and the kinetics of the HTW, it seems that polarity reversals are occurring in the retina and cortex, with wave fronts traveling at (a retinal) 7 mm per second.

In a robust display, the alternate blue bands and yellow-green reticulum arise always in tandem. In a weak display, luminance, but not chroma, is distinguishable. Triggered by photic priming and dark-adaptive transients, HTW are manifested in darkness, with a rhythmicity imposed by brainstem activity. Dark adaptation equates to deprivation of visual input, and Ermentrout and Kleinfeld, in 2001, noted that traveling electrical waves in cortex "are typically present during periods outside of stimulation."

The subjectively visible biphasic Helmholtz blue bands are most salient during physical passivity in two-minute dark-adaptive conditions imposed following outdoor activity and solar irradiation. During inactivity in the dark, the brain, this gigantic system of weakly coupled oscillators, has had the variance of frequencies reduced. After a one-minute hiatus, some critical threshold is reached and the oscillators become synchronized. In the absence of external stimuli and motor events, oscillators respond to the internal fundamental, the cardiorespiratory rhythms emitted from the brainstem, at approximately 0.25 Hz. After a two-minute interlude, the local networks systems apparently reset to their idiosyncratic rhythms, and perception of the HTW global oscillations ceases.

It is possible that these HTW are merely epi-phenomenal, serving as a visual marker of the brainstem impulses that flow throughout the CNS. After a two-minute interlude, the local networks systems reset to their idiosyncratic rhythms, overriding the vital-center emissions, and the HTW phenomenon abates.

McGuinness M., 2004: "Arnold tongues in human cardiorespiratory systems...While RSA can cause synchronization where inspiration modulates heartrate, the strongest mechanism for synchronization is CVC cardio-ventilatory coupling (CVC, where the heart is a pacemaker for respiration)."

Supposedly, bistability in opponent systems is present at all hierarchal levels in all sensorimotor modalities. Possibly, retino-cerebral dominances, sometimes patchy affairs, are compounded of multiple localized bistable (weakly coupled) neural-oscillating mechanisms. Blake considers those traveling waves as "dominance waves which may provide a new tool for investigating fundamental cortical dynamics."

Spiral Wave Dynamics in Neocortex (Topic D12) Explicit Perceptions *In Vivo*

Introduction

This report concerns the entoptic Rotating Spiral Waves (RSW) as observed and documented by the author over a period of 46 years (1962-2008). The manifestations of these state-dependent, elusive RSW were brief, emerging only during sleep-to-waking arousal epochs (*in limbo*). The images were seen only with closed lids in favorable ambient lighting – here termed the *umbral view.* The clusters of RSW emerge briefly to conscious view; their angular subtenses are estimated to be between one to four degrees, and the rotations at ten-turns per second.

Epochs of these activities commonly continued for about twenty seconds, with longevity of each active cell for up to four seconds. Ninety percent of all observed RSW cells were circular and outwardly levorotary; five percent were elliptical, appearing only as horizontal (prolate) entities. Overlapping cells were rare, and were chiefly elliptical. Observations of twin spirals were also rare, seen in counter rotations, each twin inwardly rotating. Turing patterns and scrolling were twice observed.

Summary & Findings

The subjective perceptions of spontaneous RSW are reported, with the inward spirals seen as rare and linked with rare dextro-rotations. Subjective visual resolutions of these phenomena approach those resolutions with objective VSD technology as obtained in the laboratory. When the chiralities in the cells are discernable, approximately 90 percent appear to be in levo-rotation and outwardly spiraling.

- 90 percent of all RSW cells were circular and outwardly levo-rotary.
- 5 percent were elliptical, appearing only as horizontal (prolate) entities.
- Target/Ring cells, approximately 5 percent, large and solitary, usually "popped up," and did not drift.
- Overlapping cells were rare, chiefly elliptical.
- Striate/planar waves presented as oblique or horizontal patches.
- Inward rotators, approximately 1 percent, arose from the edges of these striae, or appeared de novo.
- Twin spirals were rare, seen in counter rotations, each twin inwardly rotating.
- Ladders appeared as solitary entities or were seen within blank targets.
- Spinning tori may occur.
- Turing patterns were twice observed.
- Scrolling was twice observed.

Huang X., 2010: "Although spiral waves are ubiquitous features of nature, and have been observed in many biological systems, their existence and potential function in mammalian cerebral cortex remain uncertain."

Investigation of these waves is of special interest when the spirals arise in the visual pathway, for little is understood of spiral wave functions, if any, in the living human brain. These 10 Hz cycling waves may represent adventitious markers of ongoing excitatory neural activity, with positive or with adverse effects. According to Ermentrout (2001), "Possibly, spirals modulate our neurosyncytial communications whether of a retinal or a cortical derivation."

In 1962, I experienced some spontaneous entoptic phenomena. These I sketched and labeled as "Wheels and Standing Waves," and I preserved those contemporary sketches (fig i below), then believing them as mechanical dynamics arising from vascular turbulences. When originally encountered entoptically, these serendipitous dynamic phenomena lacked attribution, hence my correspondence in 1963 with the biophysics department at the Mayo Clinic.

At least ten papers on the Belousov-Zhabotinsky (B-Z) reaction were published in Russian before the first in English in 1967. At a conference held in Prague in 1968 on biological and biochemical oscillators, Zhabotinsky presented some of his results. The publications in 1973 finally brought recognition of the B-Z reaction to the attention of several Western chemists.

Turing Patterns: Formations in Excitable Media Entoptic Perceptions Are Here Reported

The neuroretina and the brain, as reaction-diffusion systems, appear as polar array templates, and with cross-linked lattice architecture, are deemed able to sustain both inculcate imposed patterns and the innate Turing effects.

Turing Patterns. Occasionally in 1962, and again in 2004, I have witnessed entopic displays of these spontaneous images, as described in my 1963 correspondence with Dr. E. Ackerman and colleagues in the biophysics department at the Mayo Clinic.

Turing: the Primal Code? According to M. Conrad (1982), "Once a primitive code appeared it could become more sophisticated through a multi-coding mechanism together with classical Darwinian mechanisms."

The mechanisms which—*ab initio*—generated biological patterns and structures are themselves tuned to sense the resonant natural dynamics in the external environmental patterns.

According to R. Kurtz (2003), "Physiological mechanisms of neuronal information processing have been shaped during evolution by a continual interplay between organisms and their sensory surroundings."

The Rotating Spiral Waves are found in many primitive life forms and have been deemed a putative communication mechanism. These dynamic rotors are found also in multicellular organisms and are recorded electronically in mammalian cardiac and neural tissues.

For several years I have documented the RSW, which are found subjectively perceptible. Seen entoptically as approximately 10 Hz rotations, some 90 percent are levo-rotary cells, while only 10 percent of cells are seen as inward-rotary or as dextro-rotary. These RSW may arise as epi-phenomena, indicating synaptic hyperactivity yet serving no semantic function, as some may speculate.

The neuroretina itself represents a reaction-diffusion system with cross-linked lattice, able to sustain some stable or drifting Turing effects. According to Lourenço, Babloyantz, and Hougardy (2000), "Pattern segmentation in a binary/analog world: unsupervised learning...versus memory storing, The Generation of Morphogenesis by Harmonic Resonances."

***Traveling Waves (of Helmholtz)**. These have been postulated to "scan" the CNS. As a specific search mechanism, I doubt this but found some evidence that the passage of these biphasic waves, at some four-second intervals, may enhance alternate aspects of figure/ground displays by reversing the contrasts of some persistent illusory and geometric engram images. (See Topic D11 "The Helmholtz Traveling Waves.")

***Solitons...qua...3-D Target Waves?** I identify this phenomenon with the cellular automata (See Topic B5 "Recursive Lexical Engrams (Mnemonic Recall)" in Part Two) that appear to act as lexor/cursors, commonly of a one-degree visual subtense (300 micron if in the retina). These active lexor-entities are seen only overlying the texts in those lexical engrams that may appear briefly in sleep-to-wake arousal epochs. There is in the literature no identification of these sensory agents as explicit, though their angular subtenses are implicit in Pelli's and also in Legge's earlier findings.

***Coherent Lexical Image Processings**: According to Cohen Dehaene (2007), "the acculturated semantic engram system... cultural inventions invade evolutionarily older brain circuits."

***Turing-effect patterns** and the spiral or quasiperiodic waves, which are inherent in reaction-diffusion systems, contribute to the generation and recognition of patterns and textures, and even of the text letter forms as now devised; orthography is the prime exemplar of our inculcate pattern-recognition systems.

Patterns in Excitable Media: Entoptic Perceptions

The Symmetrical Oscillating Centers, as I witnessed them entoptically in 1962, prompted my subsequent inquiries into the nature of these dynamic patterns. Naively, I had assumed that these spontaneous entoptic patterns, which resembled standing waves and vortices, might be arising from vascular vibrations in the retina. I therefore wrote to biophysicist Eugene Ackerman at the Mayo Clinic, illustrating and describing these several images as "basic patterns, and very evanescent...seen occasionally to occur spontaneously when the eye is dark-adapted and the observer is suddenly aroused from sleep. The whole field may appear as alternating grey and black circles of these varying dispositions."

These patterns I then sketched, having kept the illustrations:

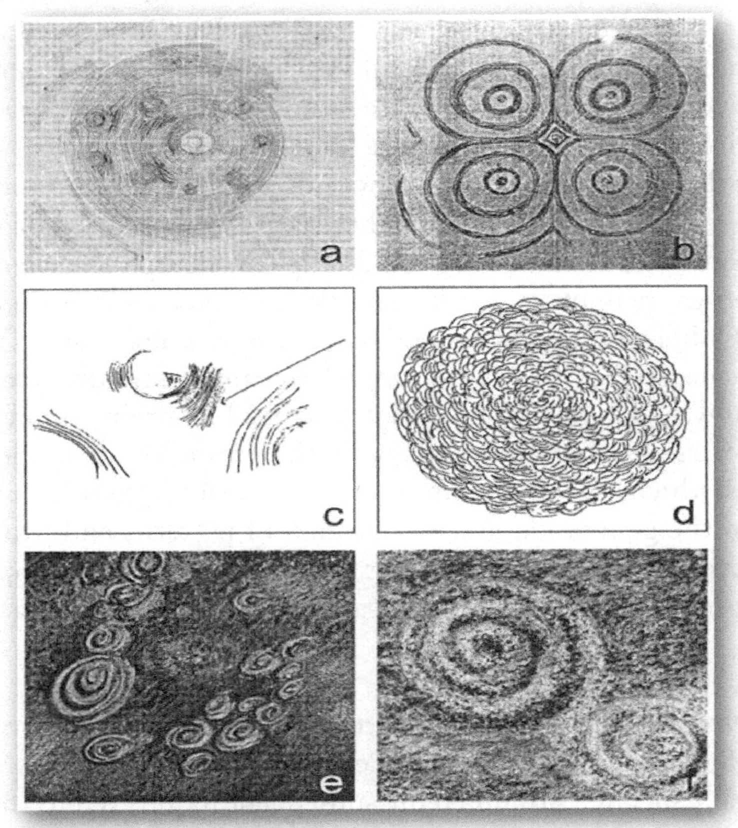

1962 Portraits of Entoptic Wave Patterns; with Current Commentary

(a) **My 1962 sketch of the Entoptic Pattern** This representation reveals a fifteen-degree subtense from fovea to disc, indicated here at margin.

(b) **Entoptic Pattern Sketch** This Turing pattern had a fifteen-degree subtense. "Very broad lines, perhaps 300 micron, centered upon four foci at 10 degrees or so from fixation of this quadrantic image, sometimes only the lower half would materialize."

Spontaneous Entoptic Images: (a) Nine small active foci lie at the intersections between the octagonal cardinal radii and two concentric circles. There as offsets depicted between adjacent inner and outer circles is a topography seen in, other geometric entoptic images. Tremulous interference patterns arise between the oscillating systems.

(b) This was seen in hypothermic conditions, this stable low-frequency mode has an oscillating center in each quadrant.

(c) Three independent foci with in-commensurate frequencies. Tremulous at arrow visually, the patterns appeared as if incomplete.

Stroboscopic Wave Patterns: (d) Concentric dark and bright "petalloid figures" multi-centric and flickering. Seen only by toggling between 30-40 Hz.

Mechanically Generated Patterns: (e) Distorted vortices result from steady oblique pressure on the globe. (f) From coronal pressure. Twin spirals, apparently dextro-rotary; dated June 1962.

Extremely fine, say 30-micron width, concentric lines centered upon the fovea, with auxiliary foci at about five and at ten degrees from fixation.

Other concentric dark and bright "petalloid figures," multicentric and flickering, were seen only by toggling a stroboscopic lamp briskly between 30–40 Hz.

September 13, 1963. Dr. Ackerman had replied that for the given frequencies my "estimated wave-lengths (of 30–300 micron) are absurdly short for plane waves of a purely shear or compressional nature...although perhaps possible with surface waves with a low interfacial tension." My attribution in 1963 of the entoptic wave phenomena to mechanical vascular turbulent patterns was thereupon abandoned.

Later interpretations (B-Z Effect) were made by January 15, 2008.

The 1962 Entoptic is now deemed a Turing pattern.

Scale: Each depicted field at approximately thirty-degree subtense. Instability was noted as a wavering at intersections of the local minor-foci within the global, centered oscillations.

The ring of the minor foci appears to be missing three members. This inner circle suggests a full complement should be that of eight nodes. The placement of the three active nodes seen in the outer circle is also consistent with the template geometry whose nodes are in octagonal lattice polar array. (See Topic D10, "Mandala: The Subjective Polar-radial Array.")

Observations made February 12, 2000. I again encountered subjective vortices while investigating the semantic hypnopompic and the recursive lexical images. (See Topic B5 "Recursive Lexical Engrams (Mnemonic Recall).") These semantic mnemonic engrams are witnessed briefly when suddenly aroused from sleep, while remaining dark-adapted in a lit ambience.

These three figures appeared simultaneously, evolving from prolate expanding cells. As expansions ceased, a spindle arose axially, terminating when its height equaled the "radius of the prolate ellipse."

These three cells operated synchronously over approximately eight seconds, apparently in a 3-D perspective view: this is inexplicable in several regards and there is visual uncertainty as to whether the spindles arose from RSW or from target cells.

Shortly thereafter, I came across graphic illustrations and a text published earlier by Gerhardt, in *Science*, in 1900. These indicated to me that my 1962 entoptic vortices were reaction-diffusion phenomena; and I further learned that the chemical B-Z diffusion reaction itself was not widely published before 1973 and the biological spiral waves in 1974. (See Winfree).

Subjective Vortices Three cells apparently in perspective.

The dynamics of the 1962 entoptics with the four-second durations and the wave dynamics are consistent with RSW data now available, specifically that of X. Huang in 2004, with the perfused neocortex mammalian tissue preparations, who then reported a four-second longevity of spiral waves with rotational frequencies of 4–14 Hz.

Investigation of these waves is of special interest when the spirals arise in the visual pathway, for little is understood of spiral wave *functions*, if any, in the living human brain.

The 10 Hz cycling waves may represent adventitious markers of ongoing excitatory neural activity, with positive or with adverse effects. Ermentrout (2001) found that "Possibly, spirals modulate our neurosyncytial communications whether of a retinal or a cortical derivation."

Entopic Turing Patterns

Turing literature, well-known to many others for decades, was unfamiliar to me before 2003.

June 5, 2002. Two episodes of geometric images were witnessed while awakening with closed lids and in a dark room illuminated by flickering TV lighting. Each of these displays had appeared to occupy an entire cognitive perceptual field. The rough sketches and notes of these brief hypnopompic phenomenal images were made rapidly and promptly while still fresh in spatial memory. The topologic lattice arrangement of these circles (as "eyes, or target cells") is evident, though slightly misaligned in the original sketch below.

The "Chevrons and Circles"

The original 2002 notes had indicated the following:

- A sudden onset of a complete (field) resembling a wallpaper of coarse structures
- Chevrons and disc/circles of equal dimensions, black and white
- Bimodal Texture and Tiles at ninety degrees only—no overlaps
- Duration twenty seconds?
- Motility, none (of the pattern) but "a large *roaming* cursor is present"

This "large roaming cursor," which moved slowly to the right across the lower third of the field, caused no disturbance of the underlying pattern.

The top-left marginal notes made September 18, 2002 suggested that a noughts-and-crosses percept might have generated this display! However, "no x were seen, only the chevrons." The dated note scribbled in the top left margin reflects a later idea that experiences of tic-tac-toe relate directly to the most basic orthographic symbols: noughts or crosses.

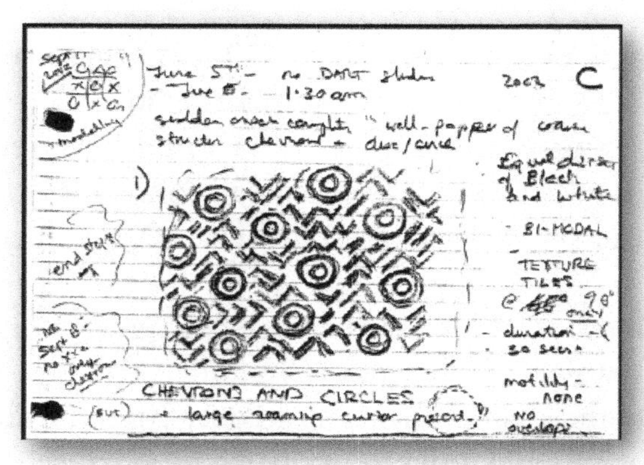

Chevrons and Circles My original written findings from 2002.

Perceptions of the 1:30 a.m. display. The overall impression of this texture pattern was visually dominated by the twelve targets (a figure-over-ground, ambiguous bias). At no point did the black rings touch the black bars of the chevrons; but some of the chevrons as here sketched are fragmented. The domain of a target cell is assumed to be demarcated by its white outer annulus.

Computer-graphic version of my 2002 sketched Turing pattern, generated in 2006.

On the June 5, 2002 display at 1:30 a.m., these twelve identical circular figures clearly resembled the so-called target cells. These circular cells appeared as interposed between geometrically antithetical chevrons; their linear configurations were unfamiliar to me other than as scintillating zigzags of migraine auras.

Global Aspect. In the overall lattice configuration, the centers of the target waves are horizontally and vertically separated by twenty units, and diagonally separated by fifteen units.

These coordinates operate globally in Cartesian and locally in the polar coordinates of each target cell.

Park W., 2007: "Interconversion between truncated Cartesian and polar expansions of images…an algorithm is applied to solve the polar-Cartesian interpolation problem." (See my Topic F17 "Visible Perception of Vestibular Signals.")

Estimated Metrics of the Local Elements

Local Features. Taking as unity the width of a linear bar, this dimension is judged equal to the width of each zone of the uniform concentric target cells. The diameter of each target cell is then seven units (or nine units if demarcated by a white zone). Thus, for the chevron-textons, given the width of their equidistant black bars also equaling one unit as does the space between them (the white bars), such a uniform metric standard suggests that this Turing pattern operates the same intrinsic wavelength across the full field manifesting in two modes, either as the chevron VVV bars, or as the circular OOO elements.

Turing pattern with proportion preservation

Ishihara S., Kaneko, K., 2007: "Turing pattern is one of the most universal mechanisms for pattern formation. In its standard model the number of stripes change with the system size, since the wavelength of the pattern is invariant."

Condat L., Van De Ville D., Forster-Heinlein B. (2008): "A new grid conversion method is proposed to resample between two 2-D periodic lattices with the same sampling density." (See Topic F17 "Visible Perception of Vestibular Signals.")

Metrics. The field of cognitive view is estimated at thirty to forty degrees of angular subtense. This measure is the cognitive *lexical* visual field, verified as a standard in which Recursive Orthographic engrams are spatially intact. The diameter of one target cell of the June 5, 2002 display may thus approximate a five-degree subtense, coincident in scale with a Lexor ii. (See Topic B5 "Recursive Lexical Engrams (Mnemonic Recall).")

The 1:30 a.m. event: The apparition of the "floating-cursor" further encourages the belief that this Turing display/map arises upon the same cerebral "texture platform" that supports the fully documented Lexical Engrams of Orthographic Texts, which are seen with such cursor/lexors. The one-degree lexor, as is consistently documented in my Topic B5, coincidently has the same magnitude as the inner zone of these Turing target cells, while Lexor ii has a five-degree subtense.

June 5, 2002. At 5:20 a.m. some hours later, the second hypnopompic display appeared in "status linbo." This was "of high resolution, with rotation of rings, five large, cellular entities of very fine grain…on a background of Eigengrau (noise, chaos) motile rotary discs dominant…the intrusion momentarily of a stick element caused no disturbance of the second display, but the contrast of the stick switched between black to white as the TV flickered."

After this second display at 5:20 a.m., these five circular figures were promptly sketched, and although labeled as rotating scroll rings, they appear depicted as spiral waves, not as target cells.

Their longevity was not recorded. The background of this 5:20 a.m. display appeared amorphous. The disposition of these five cells remained as if on the 1:30 a.m. lattice; their diameters approximately double the earlier stable target cell morphology, with dynamic spiral rotations.

M.T. Wilson and J.W. Sleigh (2007): *(Wilson M.T., Sleigh J.W., 2007)* "Turing structures: spontaneously crystallize…into random maze-like centimeter-scale spatial patterns…"

M.L. Steyn-Ross (2007): "Gap junctions mediate large-scale Turing structures in a mean-field cortex driven by sub-cortical noise."

M.L. Steyn-Ross (2009): "Interaction between the Hopf and Turing instabilities may describe the non-cognitive background or 'default' state of the brain, as observed by bold imaging."

T.I. Baker and J.D. Cowan (2009): "Spontaneous pattern formation and pinning in the primary visual cortex. Turing mechanism to generate patterns of activity."

(Migraine auras in the primary visual cortex are hemianopic. Other geometric patterns of spontaneous brain activity, including the RSW, are deemed global.)

D. Hernández and R.A. Barrio (2007): "Soliton Behaviours in a Bistable Reaction Diffusion Model…features of Turing systems: The patterns in two dimensions are particularly interesting because they can present coherent dynamics with pseudo spiral rotations…"

Such patterns might thus be witnessed in limbo (arousal from sleep) as entoptic engrams that are cycling in slow-wave sleep and manifested in a noise-driven cortex.

Zhabotinsky (2008): "Coupled and forced patterns in reaction-diffusion systems...these patterns can be travelling (e.g. spirals, concentric circles, plane waves) or stationary in space (Turing structures, standing waves)."

L. Yang, M. Dolnik, A.M. Zhabotinsky, and I.R. Epstein (2006): "Turing patterns beyond hexagons and stripes." The best known Turing patterns are composed of stripes or simple hexagonal arrangements of spots.

Ermentrout, in 2003, suggested that "increased levels of cortical excitation during the moments right after waking-arousal may play a role in manifesting spiral waves at these times."

The Turing patterns, as witnessed entoptically in 1962 and 2002, are deemed the precursors of pattern formations of nature and of orthography as the basic xx oo xxx noughts and crosses!

H.X. Hu, L. Ji, and Q.S. Li (2008): "Inward and outward spiral waves as well as inward target waves are induced by local delay feedback in a reaction-diffusion system exhibiting a Turing hexagon pattern spontaneously."

Motion Discrimination of Real or of Apparent Motion

Morrone (2000) suggests that perceptive mechanisms of for-real motion are best tuned to "radial and rotational motion." In subjective estimates of angular velocity in rotations of external displays, Barraza and Grzyacz (2002) reported that "...subjects do not estimate a unique global angular velocity, but that they perceive a non-rigid disk, with angular velocity falling inversely proportionally with radius." Seiffert, in 1998, concluded that, "The changing position of features was more readily detected than was velocity."

Apparent motions. In the laboratory, illusory virtual-motion effects are created artificially by external flickering lights, by oscillating targets, or by other devices, and the observer's neural system is thus tricked to falsely register motions. As seen in the Ternus phenomenon online, data are well documented for the detection of real and illusory translations protocols, but there is little data on rotations. According to VanRullen in 2006, "A strong temporal frequency dependence: the wagon-wheel illusion is maximal at alternation around 10 Hz, associated with changes in electroencephalogram power at approximately 13 Hz. Motion perception occurs in snapshots <100 ms in duration."

With external stimuli, when distinguished between local spiral motion detectors present in V1 and the global detectors functioning at higher levels, these higher levels were found to be subject to attentive direction (Aghdaee, 2005). According to Grossberg, in 2000, "Attentional enhancement of low contrast images also is aided by pre-attentive mechanisms." This effect pertains also to the "attentive" aspects (a mind-set) for a study of entoptic spirals. While each rotation expands the area of a cell, these "rungs" that correspond to this continued expansion are not themselves strictly countable. In attempting to enumerate the rotations in RSW, questions arise regarding the cortical motion-detection systems, which enabled those subjective perceptions of endogenous spiral dynamics and the perception of their rotations at 10 Hz. An attempt is made to compare these entoptic dynamics with the parameters reported elsewhere in the objective data obtained with exogenic rotary motions. To further this purpose, a **Circus Ternus Target** was designed and implemented. (See Topic C6 "Meandering Targets Can Reveal Edge Detection.")

Motion perceptions of these endogenous rotations. My journal of October 27, 2001 notes one entoptic RSW cell as having a *"large, clearly anti-clockwise rotation"*...This note then asked, *"If the dark and light bands are equidistant, how then can rotation be perceived?"*

This dilemma is addressed by proposing RSW motion as visually equivalent to a rotating top in a precession-wobble, a meander effect. Detection of rotation of the RSW is thus enabled by the slight shifting eccentricity in a meandering of the pivot point, which asymmetrically distracts the spiral arms, creating a discernable "hula-hoop" effect.

Meandering and the chiralities of rotations. The relationship and synchronization between the meandering and rotational phases is shown by Tung and Chan (2002). This meander dynamic was not directly observable subjectively, but the periodicity of a swing-rate could be established by an unvoiced mental counting, aided sometimes by digital (finger!) motions synchronized with the perceived entoptic rotations, apparently at around 10 Hz. (Paganini reportedly could finger his violin at 12 Hz.) The cycles themselves could be counted off by attention to the slight distraction between adjacent quasi-parallel rings of an axially unstable or meandering cell.

Displacement of the features by the wobble may allow these subjective perceptions of RSW cycling. The rotations and chiralities is clearly perceptible in less than one-third of all observed cells, but when it is detectable, anti-clockwise levo-rotation is noted in about 90 percent of this population.

A chiral bias is little noted in the earlier literature. X. Huang states that cells arising in the same location present the same chirality as their predecessors. N.L. Smith (2004) verifies at a molecular level that the handedness expected for helical waves was identified with "...precession in the opposite sense to the rotational flow around the vortices."

Angular velocity may be constant in small, stable, rigidly rotating circular spirals at 360 degrees in one-tenth of a second (Winfree, 1998). In the horizontally elliptical spirals, always prolate, the linear ratio of their orthogonal axes was commonly 0.5.

Linear velocities. For certain traveling waves such as the Helmholtz Traveling Waves, the global motions appear subjectively as circular, elliptical, or hyperbolic wave fronts traveling across the visual field. These bands travel either centrifugally *or* centripetally, as determined by the locations and asymmetrical sites selected to be triggered by photic radiation or by focal pressure applied to either globe; thus, the fronts can be manipulated to be asymmetric and to be briefly out-of-phase as between the two eyes. The HTW bands move across the full visual field/retinal arena in approximately three seconds.

For a spiral disturbance racing along the outer arms of a large 15 mm spiral rotating at 10 Hz, the calculated linear velocity might reach four centimeters per second around this "circular track." The slowest observed rotation, documented on December 3, 2004, was only 3 Hz. This cell collapsed, first to a cog-wheel circle and then to a smaller granular interior: a unique event.

Inward Rotating Spirals, Dynamical Transitions, and Other Rare Events Chirality Biases, Twins and Solitaries, and Toric Helical Cells?

Reportedly, in the chemical B-Z reactions in biological preparations and in cardiology, the chirality is indifferent. It is therefore remarkable in the entoptic manifestations that levorotation prevails in over 90 percent of the cells where their chirality can be discerned. In large artificial cell assemblies, H. Skødt (2003) found "anti-spirals extremely rare." J. Davidsen (2004) discusses "…the formation of various spiral-wave attractors, in particular pairs of spirals in which one spiral acts as a source and a second as a sink—the latter similar to an 'anti-spiral.'" This I have not yet experienced as a subjective phenomenon. In mathematical models, Tsyganov and Biktashev (2004) showed that the angle of collision, curvature, and period of conflicting waves determined the outcomes; this dynamic I have observed and documented.

Twin spirals. As witnessed in a human retina, the counter-rotating inward-flowing twins have rarely appeared. The dimensions of associated striate patches apparently determine the final diameter and the granularity of their spiral progeny. Counter-rotating twins, apparently unrelated to striae, have also been seen with mutual annihilation in a visible starburst (June 15, 2003).

Chiralities were not distinguished subjectively before October 29, 2001, yet an apparent bi-chirality was depicted in 1962. Thereafter, more recent journal entries indicate that inward rotators were frequently derived from oblique striae that can create a pair of rotors, both inward-rotating. According to Winfree (1998), "Gradient of phase is created by the recent passage of a travelling wave, and the gradient of stimulus intensity is generated by creating a temporary stimuli on the medium … this mechanism creates a pair of mirror image rotors." Those cells generated by the striae were documented as levo-rotary (though in Figure 2, B-2 are portrayed as dextro-chiral, which are rare.) As stated by H.S. Hu in 2008, "Compared with the inward spirals, the outward spiral waves usually possess longer wavelength and exhibit larger amplitude relaxation oscillations."

Some Brief, Dated Entries on the Inward Rotations

It was noted in my sixteen recently recorded cases of inward rotations that the rare dextro-rotations coincided with inward rotation in at least six instances. Although this series inevitably was small, this ratio seems significant since circular spiral cells are 90 percent levo-rotary, and the observed, overall incidence of inward rotation is perhaps one in two hundred cells.

The Inwardly Rotating (Biological) Spiral Waves

Inward spirals were reported in nonorganic excitable media by Vanag in 2002, and by Gong in 2003, but apparently this inward dynamic had not been noted in biological tissues, nor has the in vivo low incidence (greater than 10 percent) of apparent dextral-spin been reported. In this present experiential report, those two features have appeared linked, the *rare inward-spiralings* being linked 50 percent to the *rare* dextro-rotators as evident in vivo.

When, in 1962, I had serendipitously encountered RSW, these subjective phenomena then lacked attribution. These spontaneous cortical activities become evident as the dark-adapted observer stirs from sleep, and only then if the eyes remain closed but illuminated—termed an *umbral view*. The clusters of rotating spiral waves (RSW) which thus arise (arguably either in the brain or the retina) emerge briefly to conscious view, their angular subtenses from one to five degrees, and the rotations are estimated at ten turns per second. Epochs of these activities continue commonly for about twenty seconds, with longevity of each cell for up to four seconds.

The following data accrued from the several thousands of such cells I witnessed over more recent years.

- 90 percent of all RSW cells were circular and outwardly levo-rotary.
- 5 percent were elliptical, appearing only as horizontal (prolate) entities.

- Target/Ring cells approximately 5 percent, large and solitary, usually "popped up," did not drift.
- Overlapping cells were rare, chiefly elliptical.
- Striate/planar waves presented as oblique or horizontal patches.
- Inward rotators, approximately 1 percent, arose from the edges of these striae, or appeared de novo.
- Twin spirals were rare, seen in counter rotations, each twin inwardly rotating.
- Ladders appeared as solitary entities, or were seen within blank target cells.
- Spinning tori may occur.
- Turing patterns were twice observed.
- Scrolling was twice observed.

Mirror Twins. Winfree, 1998, noted "*the recent passage of a travelling wave, and the gradient of stimulus intensity is generated by creating a temporary stimuli on the medium…this mechanism…creates a pair of mirror image rotors.*" In these reported *subjective* views, each such paired twin has been inward rotating and of opposite chirality of the RSW arising de novo, 90 percent being levo-rotary, and most rotating outwardly.

January 6, 2009. Only in a meandering cell can the RSW dynamics be discerned as inward or outward spiraling, chirality detectable, and with the rotation rate estimable.

Chirality commonly appears uniformly as dextro-rotary in inorganic and in biological substances.

According to Hu X. Huang (2008), "Inward and outward spiral waves as well as inward target waves are induced by local delay feedback in a reaction-diffusion system exhibiting a Turing hexagon pattern spontaneously."

Inward rotating spiral waves were reported in nonorganic excitable media by Vanag in 2002 and by Gong in 2003. Bard

Ermentrout, in 2003, responded that he "had seen these patterns, but had not at that time quantified them."

Winfree, in 1998, had noted that "the recent passage of a traveling wave, and the gradient of stimulus intensity is generated by creating a temporary stimuli on the medium…this mechanism then…creates a pair of mirror-image rotors."

Helmholtz in 1924 had described "subjective blue-band traveling waves of ocular origin," and H.R.Wilson in 2001 reported "visual perception of cortical traveling waves."

Arising in the human brain, the patterns of ionic perturbations, such as auras, are familiar to migraineurs. The voltage-sensitive neurons in the brain are also responsive to RSW oscillations and are hence subjectively perceptible, as also are the Helmholtz traveling waves. (See my Topic C6, "Meandering Targets Can Reveal Edge Detection"—a hula-hoop phenomenon.)

Voltage sensitive dyes (VSD) enable photographic images of rotating spiral waves in many excitable biological media. Such data was critically studied first in the syncytial myocardium, more recently in neural tissue, in vitro, and here, *in vivo.*

Chromatic-Entoptics (The E Series)

Ocular Blood Flow and the Level of Neuro-Retinal Activity Are Both Maximal at an Imposed 40 Hz Manifested Subjectively by a "Chromatic-Entoptic Annulus"

According to C. Riva, in 2005, "The linkage at ~40 Hz between maximal neuro-retinal activity and ocular perfusion levels now appears incontrovertible."

"Purple Entoptic Annulus" phenomenon is visually evident at 40 Hz, offers a simple means to assess acute fluctuations in ocular neurometabolism, and indicates the prompt auto-regulation of ocular perfusion levels (IXP, 2000).

The sources of this phenomenon, its topography, and the mechanisms that enable its subjective perception are of interest and have possible clinical application. Limited testing in a clinical setting distinguished low-tension glaucoma from ocular hypertension.

Pathology delays the onset of this phenomenal annulus. By reversing the initial testing order of the two eyes, one can then differentiate the performances between the two eyes and thus distinguish any unequal performances as being attributable either to monocular pathology or simply to the asymmetrical prior light adaptation.

Aging, with diminishing ocular circulation, contributes significantly to prolonged delayed onset of the marker, often to forty seconds. According to Cristini, Forlani, and Scardovi in "Choroisal Circulation in Glaucoma," 1962, there is a "40% decrease seen in choroidal thickness by age 75."

Anomalies. Interestingly the dimensions, morphology, topology, bandwidth, and kinetics of this purple phenomenon have been attested to by an albino patient, in 1964, and also by two severely red-green color-blind individuals in 1966 and in 2002.

Diverse views are offered regarding pathological vectors:

L.R. Band (2009): "In glaucoma, an elevated intraocular pressure causes a progressive loss of retinal ganglion *cells…resulting in the optic neuropathy.*"

Levin (2007): "Glaucoma is an optic neuropathy in which the optic nerve axons are damaged, *resulting in the death of retinal ganglion cells.*"

N. Gupta, Y.H. Yücel (2003): "There is evidence that glaucomatous damage extends from retinal ganglion cells to *vision centers in the brain.*"

J. Flammer (2008): "The increase of IOP in POAG is due an increased resistance of aqueous outflow through the trabecular meshwork (TM). Even less is known about the pathogenesis of glaucomatous optic neuropathy."

Schmetterer (2007): "Primary Open Angle Glaucomas show a large diurnal fluctuation of ocular blood flow supporting the hypothesis that POAG is associated with *vascular dysregulation.*"

According to Caprioli, in 2010, "The physiologic nocturnal-dip in blood pressure is protective against systemic end-organ damage, but its effects on glaucoma are not well elaborated or understood."

According to D. Schmidl and G. Garhofer (2010), "The complex interaction between ocular perfusion pressure and ocular blood flow—Relevance for Reduction of IOP by pharmacological intervention improves optic nerve head blood flow regulation independently of an ocular vasodilator effect."

Changes in the ocular hemodynamics result from changes in the intraocular pressure...and variations in the intraocular pressure accrue *from* the changes in ocular hemodynamics...both contributing to the optic neuropathy?

An Entoptic Purple Ring Phenonmenon

Using a Xenon Strobotac, 1531, toggled between 10–60 Hz, and viewed through a diffusing screen, I was then surprised to witness this purple ring phenomenon and to find it maximal at 40 Hz.

Eccentric Viewing of the Diffusing Screen

Scale: Each figure occupies fifteen degrees subtense
A xenon strobe lamp is toggled across a 15 to 60 Hz bandwidth
and is viewed through a diffusing screen.

With eccentric fixations "a purple donut" may be seen "floating about" on the diffuser screen. When the gaze toward the light source is shifted a few degrees off-center in any direction, the perception of the ring changes, as here diagrammed, and is evident only between the fifteen to five isopters. Shown here on further right gaze; only an isolated purple disc is seen. On increased eccentricity of gaze, no purple areas remain visible.

Coincident with ocular blood flows, both are maximal at 40 Hz. These initial studies of this "EP40" were made in 1962, when I had participated in a UCLA Glaucoma Tonographic project. While attempting to measure capillary retinal flow rates with a stroboscope toggled between 5 and 60 Hz, I was surprised to encounter a purple ring effect.

Unaware at that time of the earlier substantive report of an Entoptic Purple Phenomenon (Shurcliff in 1959) with **spectral studies,** I later found Welpe' s **Frequency-Modulated** study, as published in 1978. (Neither report then related the phenomena to considerations or explorations of ocular perfusion levels.)

However, my 1962 phenomena were verified by some fifty participants, including children, college students, and residents of a retirement community. Children younger than six years also participated successfully, greeting "a purple donut" at approximately 40 Hz.

The xenon stroboscopic flash lamp was toggled between 10 to 70 Hz, and at a single flash duration of 1.2 microseconds, the lamp observed binocularly through a diffuser.

With natural pupils and with random preadaptations to light, we measured the onset (latency in seconds) and the evolution of this entoptic-purple corona phenomenon at frequencies ranging from 65–20 Hz. Delays varied from zero to over forty seconds, yet regardless of pupillary diameters and levels of flux, the minimal latency remained at 40 Hz, plus or minus two, and the absence of the hue at the fovea centralis was generally remarked upon.

This noninvasive, low-technology tool might invite further study for research purposes: to measure neurometabolic-vasomotor responsivity to acute photic stress and to record the response to therapeutic interventions in cases of circulatory pathology and vaso-activity of glaucoma medications.

The EP@40: Clinical Applications

This phenomenon offers a noninvasive means for assessing the acute changes of ocular blood flow in response to imposed photic stresses. The entopic marker EP@ 40 can be manifested by flicker, promptly with healthy eyes and delayed or absent with pathology or in dark-adapted eyes. According to C. Riva (2005), "The linkage at ~ 40 Hz. between maximal neuro-retinal activity and ocular perfusion levels now appears incontrovertible."

Increased horizontal ganglion synaptic activity at 40 Hz was reported (Smith V.C., 2001).

Instructions to the Patient

The "Purple Donut" Test

You will be asked to watch flickering light on a screen. After a few moments, a "ghostly purple donut" will appear to be **on** this screen, but this is an image created within the eye itself. A prompt appearance of the purple donut indicates:

1) that your eye has a healthy blood supply and a "good chemistry."
2) that you, the participant, are already well adapted to bright light.

A delayed appearance of the purple donut is normal if you have been sitting quietly in the dark for some minutes, but during the ten-minute light test, a healthy eye adjusts and the onsets accelerate. Failure to adapt might indicate a circulation problem for further study.

The test takes three minutes or less for each eye.

The purple donut vanishes instantly when the light is turned off.

This test has been carried out in over seventy cases; there are no adverse effects.

The results of your test will be explained to you.

"A Purple Response Index"

Performances may be scored by combining two measures. In the denominator, the delay at 40 Hz in seconds; and as the numerator, the bandwidth in Hz; this ratio can represent the Purple Response Index. An "instantaneous" perception of the phenomenon thus rates a nominal one- second in the denominator. Delays exceeding forty-five seconds often coincide with a bandwidth so narrow as to be barely perceptible to the subject; such a "poor" performance receives a nominal unity in the numerator. The highest attained score has been forty-five.

Purple: A Nonspectral Color (Topic E13)

The literature from 1823 refers to "entoptic violet and yellow-green flecks" seen with intermittent light.

Gebhard, in 1943, mentions a "ubiquitous violet cloud floating about" during intermittent illumination; this hue was also reported by Brown and Gebhard in 1947, by Smythies in 1950, and by Shurcliff in 1959. In addition to a "reddish-blue" entity, Shurcliff noted greenish-yellow "antipathic" phenomena, whose kinetics differed from those of the purple mode Welpe, in 1978, studied a "violet effect…originating in the retina and visible from 28–43 Hz. peaking at 40 Hz. +/-2 Hz." This is also the maximal oscillatory activity of second-order retinal neurons identified by Smith in 2001.

The phenomenology of this entoptic purple ring is detailed in the following Topic E14, which addresses the relationship with photic stimuli, its peak with neurovascular coupling at 40 Hz and with enhanced ocular perfusion.

According to C. Riva (2005), "The linkage at ~40 Hz. between maximal neuro-retinal activity and ocular perfusion levels now appears incontrovertible."

This entoptic purple phenomenon is evidently universally experienced, as indicated by the ninety-five participants in Shurcliff's 1959 spectral studies and by the thirty-five participants in Welpe's 1978 frequency-modulated study. Shurcliff found no clear relation between the wavelength of the stimulus and the apparent hues of his induced phenomena; Welpe stated the color was most intense "between 28–43 Hz. and that the color... the violet effect...was independent of the color of the stimulus." Shurcliff suggested that the mechanisms related to "anomalous ratios of radiation-activated switches in parallel...[and] represent cone-functions."

This opponent phenomenon is identified also in the global Helmholtz traveling wave where centrifugal waves, the purple hue (R+B) bands alternate with the yellow-green bands. (Yellow hue = Red + Green.)

Purple is nonspectral; its opponent is yellow-green. These sections (E13, E14, E15, and E16) review the triggers, domains, boundaries, and local textures (granularity) of the phenomenal colors, the mechanics of bistable hues, and the highly significant role of a preadaptation by the ocular blood flow in response to light.

According to P. Maquet, in 2006), "Daytime light exposure dynamically enhances brain responses."

The Dominant Purple Manifestations versus the Subjugate Yellow-Green Perceptions (Topic E14)

The subjective findings date from 1964, with insights from 2004.

This unsought purple phenomenon was found maximally evident at an imposed 40 Hz. This was then verified by some fifty participants, including college students, the residents of a retirement community, and children who exclaimed "the purple donut!" Replicated more recently (2001–2008), and using the identical xenon lamp (the surviving old Strobotac 1531), the earlier data have withstood the test of time, now over four decades. The visual onsets can vary from a nominal "zero" to almost one minute. The chromatic annulus appears most immediate and salient at an imposed 40 Hz. It emerges to view between the fifteen-degree and five-degree isopters, a region where the majority of retinal ganglionic cells synapse with bipolar and horizontal cells.

The four or five bands of alternating "pixels" appear to travel for about seven degrees, vectored peripherally from the open cusp and being redirected as foveal refixations of the light source, and rotate the orientation of the crescent. See "Ionic Waves and Pixels" Topic E16.

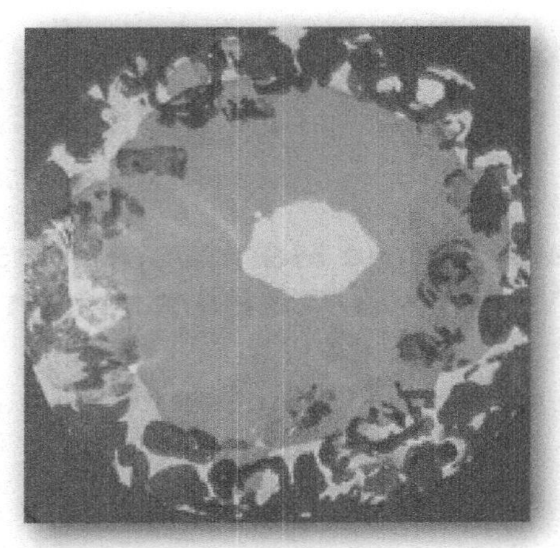

Foveal zone The foveal zone, an oval of five-degree subtense, generally remains clear of the colors that flood the parafoveal field. The outer margins of the annulus seen at 40 Hz may fluctuate briskly as yellow/white/black globules or lobules. The foveal zone is a pale purple, represented here in grey.

Eccentric viewing of the light source converts the annular configuration to a crescent, its opening in the direction of the global vector, as diagrammed below in four oculomotor versions. (Figures are not to scale).

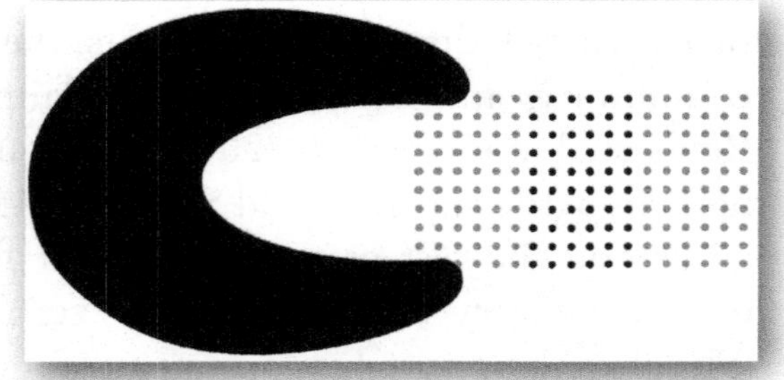

An enlargement of the right portion of the adjacent image, annular configuration. The color is purple, represented here in black and white.

Illusory motility of these chromatic particles is attributed to firing of adjacent voltage-sensitive receptors; these chains are sequentially activated by local ionic perturbations of their fields.

The Spectral and Frequency Studies. The entoptic purple phenomenon was experienced by the ninety-five participants in Shurcliff's 1959 spectral studies and by the thirty-five participants in Welpe's 1978 frequency-modulated study. Shurcliff had found no clear relation between the wavelength of the stimulus and the apparent hues of the phenomena; Welpe stated the color was most intense "between 28–43 Hz. and that the color...the violet effect...was independent of the color of the stimulus."

This present report examines two other parameters: adaptation to steady light prior to the testing and the onset delay during the exposure to the online test flicker.

Onsets delayed over forty seconds were commonly recorded with dark-adapted eyes. In any one individual, these varied from a nominal "zero" up to forty seconds of imposed flicker, predictably dependent upon current physiological status. Onset is hastened by prior physical activity and by light adaptation and is delayed severalfold or may be absent, reflecting passivity or somnolence, aging or pathology. Limited testing in a clinical setting distinguished low-tension glaucoma from ocular hypertension.

Increased metabolic activity, in the pigment epithelium, the photoreceptors and neuro-retina, demands increased ocular blood flow, notably in the choroidal circulation. The following protocols address physiological variables obtained with healthy eyes in 1964, and in 2000.

- Pupil status is controlled or monitored during the adapting and during the testing.
- Levels of personal physical activity and vigilance are recorded.
- Preadaptation noted to light or to dark, with natural pupils.

- Fast tracking, with onsets measured only at the standard 40 Hz.
- Parameters of the angular dimensions of the central hypo-chromatic ellipse.

Process: A stroboscopic xenon lamp is viewed at arm's length; a diffuser is placed on the instrument, or is worn as spectacles, or held as a shield. When gaze is directed centrally, the ring is seen. With off-centered visual alignment, a crescent appears, and when further offset, a disc of diminishing dimensions appears. (See Figure 2 above.)

Intrasubject testing with flicker shows that, for a given individual, the immediately prior photic-adaptation plus the current ambient light and/or flicker together determine a predictable phenomenal onset.

In a general population, the finding of prolonged delayed onset or a complete failure to manifest the flicker effect may indicate ocular pathology, somnolence, poor health, vasculopathies, or aging. The source of this purple-green entoptic phenomenon, its topography, and the mechanisms that enable its subjective perception are of interest. This phenomenon also offers possible clinical applications for an inexpensive, noninvasive measure of acute changes in ocular perfusion in response to imposed visual tasks to luminance or following acute therapeutic interventions.

Limited testing in a clinical setting distinguished low-tension glaucoma from ocular hypertension.

Z.Q. Yin (1997): "Widespread Choroidal Insufficiency in Primary Open-angle Glaucoma."

J. Flammer (2001): "Relationship between Ocular Perfusion Pressure and Retrobulbar Blood Flow in Patients with Glaucoma with Progressive Damage."

The Prompt Ocular-Perfusion Marker Offers Applications in Glaucoma Management (Topic E15)

Estimates

How can ocular blood flow be measured? According to L. Schmetterer (2006), "There is still no gold standard for the evaluation of blood flow in humans available; and sophisticated and expensive equipment is required."

Literature 1975–2008

The Ocular Blood Flow in Glaucoma

Autoregulation of ocular blood flow to the choroid was at one time contested.

M.F. Armaly and M. Araki (1975): "Effect of ocular pressure on choroidal circulation in the cat and Rhesus monkey. Vascular bed of the choroid in these experimental animals is a passive one without evidence of active regulation."

A. Bill (1975): "Autoregulation of the blood flow is intermediate in the ciliary body and very poor or absent in the choroid..."

A. Bill and S.F. Nilsson (1985): "Reductions in perfusion pressure, caused by increments in intraocular pressure, or reductions in mean arterial pressure reduce the blood flow in the choroid. In the retina, there are efficient autoregulatory mechanisms that prevent changes in flow within a wide range of perfusion pressures."

L. Schmetterer (2005): "Effects of moderate changes in intraocular pressure on ocular hemodynamics in patients with primary open-angle glaucoma and healthy controls. The present study does not provide evidence for altered autoregulation in patients with POAG during a moderate increase in IOP. However, these results do not necessarily contradict the concept of vascular dysregulation in glaucoma."

L. Schmetterer (2007): "The data indicate that the choroid regulates its blood flow better during exercise-induced changes in MAP than during an experimental increase in IOP."

P. Galambos (2006): "Compromised autoregulatory control of ocular hemodynamics in glaucoma patients after postural change. Measuring and interpreting ocular blood flow and metabolism in glaucoma."

A. Harris (2008): "The imaging technologies most commonly used to investigate ocular blood flow, include color Doppler imaging, confocal scanning laser ophthalmoscopic angiography with fluorescein and indocyanine green dye, Canon laser blood flowmetry, scanning laser Doppler flowmetry, and retinal photographic oximetry. Each imaging technique's ability to define vascular function and reveal pathology is discussed as are limitations inherent to each technology."

Using a xenon flash stroboscope, in 1962 the following structures were identified subjectively:

At 12 f.p.s. columns of corpuscles in the peri-foveal retinal capillaries

At 25 f.p.s. hexagons of 20 micron in diameter

At 30 f.p.s. retinal vasculature down to and including the capillary vessels

1967. Auto-regulation. With digital compressions of the globe, the imposed amblyopia begins peripherally, and once apparent, sweeps steadily centripetally, the fovea being the last affected. After six seconds of judicious, steadily sustained pressure, the amaurosis recedes from the center, and the apparent brightness of the visual field is restored overall. This response is fatigueable, but could be repeated after a two-minute respite. These findings suggest that the posterior polar circulation is less readily impeded than is the anterior retinal circulation and that vascular auto-regulations are entoptically detectable. (See Topic E13 "Purple: A Nonspectral Color.")

With a *Baillert Opthalmodynamometer*, compression of the globe sustained at forty millimeters enables the retinal capillary chains to be seen; at sixty millimeters of pressure, arteriolar pulsations and hexagons may be seen. At around forty millimeters of pressure, the slowed corpuscular flow in the retinal capillary network is seen. Corpuscular columns and their movement are not perceptible outside of the twelve-degree isopter and are detected by the retina only with illumination and angular velocity within a certain range. Kato (1951) found the rate of capillary flow measured entoptically to be seventy-seven millimeters per second. Retinal corpuscular columns are seen entoptically just prior to the retinal arteriolar systolic pulse-wave crest when the flow appears most retarded in the capillaries.

At about sixty millimeters of pressure, the pulse wave in the arteriolar retinal tree is visible as a dark shadow, which clearly arises from the blind spot and flashes across some two-thirds of the visual field. At mid-diastolic levels when the impedance in the retinal circulation exceeds the zero-flow intercept for diastolic pressure, the following picture may be noted: amblyopia of the field with mosaic pattern pinwheel vortices in the paramacular area and hexagons.

Subjectivity has been progressively excluded from the practice of science, leaving an essentially secular analytical paradigm (Jahn R.G., Dunne B.J., 2007).

April 1962. On suddenly releasing pressure, a flash of segmented retinal flow was apparent, possibly indicating the partial obstructions at arterio-venous crossings. See author's subsequent branch vein occlusion (2003) below, resolved in three months, which reoccurred in November 2011 at the identical location: site of a unique anterior V to posterior A crossing; all others are A to V crossings. See drawing below.

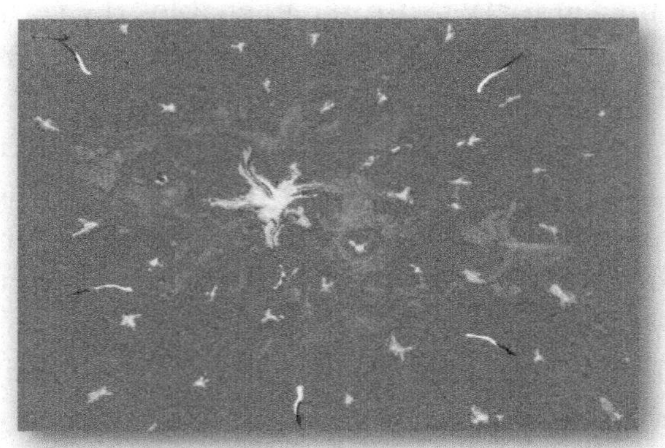

The sudden release from an IOP imposed at the mid-diastolic level demonstrates an instantaneous replenishment of retinal vascular flow. Segments of black/white flashes appear, initially peripherally, and sequencing centripetally along the vascular tree toward the blind spot. This phenomenon can be attributed to the suddenly enhanced flow in the retinal vessels and a "mechanical jolt" causing a local phosphene to arise from the immediately contiguous percipient elements (c.f. pressure phosphenes, see G. Oster). The locus of these segments may represent arterio-venous crossings and predict sites for eventual branch vain occlusions

Photic After-Images, Quasi Scotomas (Topic E16) "Ionic Waves and Pixels"

Scotoma (Latin *scutum: a shield* – Greek *scotos: darkness*). This term may refer to any localized anomaly in the perception of hue or motion or luminosity in a circumscribed area. These defects can be monitored in the absence of or with the addition of further steady visual input, by use of flicker, with variants of spectra, or by use of flux, made to both or to either eye. They resolve by attrition at their borders, marked by red rims.

After-images are considered as temporary, relative scotomas; their resolution dynamics may reveal sequential resets. The shrinking area of a central luminous core of a reset after-image is bordered by a narrow red rim, as was reported by Stamper in a laser study in 2000. This red border is comprised of uniform and linearly disposed granules: "pixels." These are not engrams.

Foveal scotomas from both candlelight and from solar specular reflections have circumscribed but shrinking zones within a two- to four-degree perimeter. As seen in the dark, the decaying border of the after-image (the black outer margin) has a narrow red rim with obviously granular margins. The granularity of this process is evident at the boundary of the hues; here the granules are seen as uniform in size with rounded contours, and they appear alternately aligned in ranks and files.

Schematics of Foveal Zones: Bezolt-Brucke Phenonmenon

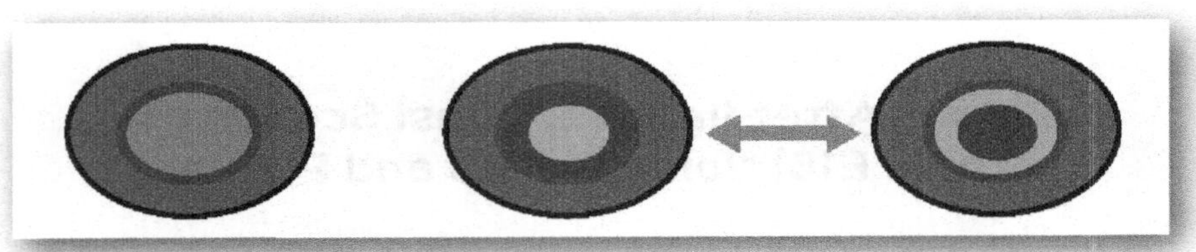

Inside this red rim is a luminous yellow area. This core at first sight is apparently featureless.

The hue of this core area is then subjected to flux-dependent chroma reversals.

The higher flux generates a pink core with a green halo. In the image, pink is shown in light grey and green is shown in darker grey. On doubling the distance from the light source, these hues reverse, green now at the core with a pink peripheral halo. These hues do not merge or spontaneously alternate. The distinct boundaries between the two hues are most easily observed when a linear solar scotoma has been imposed and a scrupulously careful observation is then made at the critical instability flux level, the switch point. Under most conditions, bistable green and pink have predominated, reminiscent of second-order subtraction colors. The stable red rim, shown as a grey ring in the image, contracts as the scotoma area is erased.

After-images are modified by changes in flux, not flicker frequency:

1) As seen in bright steady light.
2) As seen with closed lit lids (red, mesopic).
3) As seen with full spectrum flicker at 10 Hz (by Strobotac), the finer structures of the luminous core can be discerned;

their granular scale seems at the limit of separable (non-vernier) perception.

4) At a steady flicker and varying the distance from the light source (rocking back and forth: a toggle effect), chromal bistability can be identified, sustained, and monitored for some minutes. The distinct boundaries between the two hues are most easily observed when a linear solar after-image scotoma has been imposed, and a scrupulously careful observation is then made at the critical instability flux level: the switch point.

Spillmann L., 2006: "…an afterimage resulting from a strong foveal light flash can be made to pulsate by luminance modulation of a surrounding annulus as far as 8 degrees away."

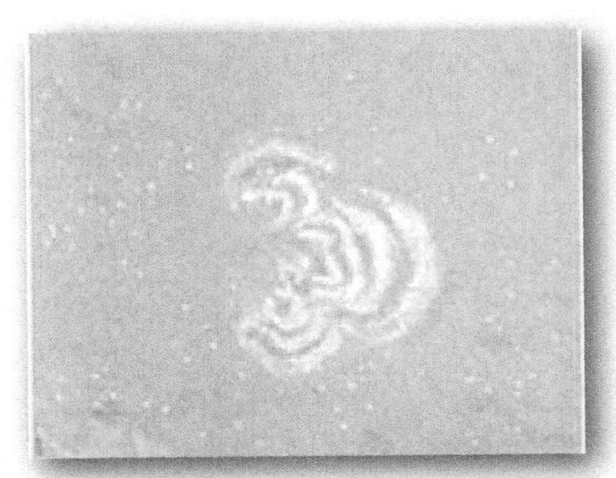

The four or five bands of alternating "pixels" *appear* to travel for about seven degrees from the lips of the purple crescent at 40 Hz.

Refer to Topic E14, "A Purple-Green Entoptic Manifestation" for the chromatic data.

The granularity is observed with an imposed 40 Hz together with the incidental advent of traveling waves. Bistability of chromal-luminance perceptions with adjacent hues are reversed by changes in the flux at distance, not by the frequency changes of flicker. Pointillistic uniform granularity arises within the foveal scotomatous zones. Small purple "islands" emerge from a "green-border (mach) band" contiguous to the purple "coastline" enfolding the cusps of a "crescent C" configuration.

Glaucoma and Related Vasual Studies (Appendix)

In 1962, an aqueous regulation study was initiated during my participation in a UCLA glaucoma research project on ocular outflow.

Incidental to a UCLA tonographic glaucoma study in 1962, I encountered the "Entoptic Purple Annulus" (EPA40).

Pertinent Questions

Do moderate changes in the ocular hemodynamics result from changes in intraocular pressure, and/or do the changes in the intraocular pressure accrue from changes in ocular hemodynamics?

J. Flammer (2007): "What is the present pathogenetic concept of glaucomatous optic neuropathy? Oxygen and blood flow: players in the pathogenesis of glaucoma."

M.C. Grieshaber and J. Flammer (2005): "Therapeutically, both an intraocular pressure reduction and an improvement of the ocular blood flow might be considered."

L. Schmetter (2007): "The choroidal blood flow in the subfoveal region of the fundus exceeds that of the retinal artery...by an order of magnitude."

L. Schmetter (2009): "The baseline values fundus pulsation amplitude (FPA) and optic nerve head (ONH) blood flow were lower in glaucoma patients as compared with age-matched and gender-matched healthy controls…these results do not necessarily contradict the concept of vascular etiology."

R.N. Weinreb (2009): "Correlation among choroidal, parapapillary, and retrobulbar vascular parameters in glaucoma…further evidence of vascular dysregulation in POAG."

The presentation in this EPA40 Series describes a noninvasive and clinically applicable means for assessing the ocular blood flow changes made in response to briefly imposed photic stresses or to therapeutic interventions.

With flicker at 40 Hz, a physiological purple entopic annulus manifests itself promptly in healthy eyes; it is delayed or absent in the impaired eyes. The critical state-dependency issues and a response indexing measure are addressed.

The genesis of neuropathy in open-angle "idiopathic glaucoma" is attributed to impaired ocular perfusion.

These data support the hypothesis that POAG is initiated or associated with vascular deregulation.

Ocular Pathologies and the Co-Morbidities Systemic Diseases and Senescence

Background: The project design by Robert Christensen was to make statistical use of the tonographic data collected for linked clinical and histological investigation of the earliest changes in Primary Open Angle Glaucoma: POAG. To this end, all the selected cases were military veterans with serious medical conditions but lacking evident ocular pathology. Such patients with no history of glaucoma were recruited at a VA hospital in California.

With an average age of sixty years, this population had intraocular pressures averaging 13 mm Hg Schiotz.

It was anticipated with the data from over the one thousand tonograms obtained in the span of several years that eventually some autopsy material might become available for microscopy (a long shot, and unachieved.) However, since most of these patients presented with multiple and well-documented systemic disorders, their tonographic data might also be processed statistically.

It was hoped to distinguish the impact of aging, diabetes, hypertension, pulmonary cardiac disorders, and systemic drugs upon parameters of the standardized tomograms.

These selected VA patients were not challenged with the Homatropine. Protocol was designed by Robert Christensen, UCLA; conclusions published thereafter were restricted to the "Homatropine Hydrobromide: Effect of Topical Administration upon the Intraocular Pressure and Aqueous Facility Values of Normal and Chronic Simple Glaucomatous Eyes" (Christensen R.E., Pearce I., *Arch Ophthalmol.*, 1963).

Consensual Aqueous Outflow Responses to Monocularly Imposed Pressure

Aqueous outflow studies were made in 1962–19633 at the Veterans Administration Hospital at Long Beach, California. A consensual IOP response indicates an intact interocular autonomic-neurosystem transmission.

Consensual Interocular Effects. Homatropine partially blocked the consensual hypotensive effect that occurs in the second eye during tonography.

Consensual Ophthalmotonic Responses. The consensual tonographic effect in normal and glaucomatous adults. According to S.M. Drance and R.L. Wiggins, in 1967, "The second eyes, yet untouched, regularly demonstrate a lower IOP."

Ocular Tonography. Although now obsolete, Schiotz tonometry has been long since replaced by Goldman Applanation tonometry for measurement of intraocular pressure: two minutes of the Schiotz pressure was thus applied monocularly and then repeated on the second eye.

In this phalanx of patients, their mean age was sixty-five years. The mean "opening pressure" IOP was 13.5 mm Hg. for the first eye and 13.0 mm Hg. in the second *slaved* eye with its consensual effect promptly tested at two minutes. The mean aqueous outflow and the Po/c values calculated. Consensual responses to monocular stimuli include the pupillary, accommodative ciliary, the aqueous outflow route, and the vascular perfusion responses.

Unreported finding from this earlier VA study: I noted prior vagotomies for peptic ulcers in thirty-nine cases had diminished consensual responses; while in the cases with medically treated peptic ulcers, the consensual responses were normal.

Vagotomies. A treatment then of complicated post-bulbar peptic ulcers.

Tonography by indentation increases the hydrostatic pressure head and the discharge of effluent into Schlem's canal. Tonography by raising the IOP impedes intraocular perfusion and diminishes the net volume, or the secretion of aqueous humor. Tonography is a simple provocative stress test.

Comorbitities. Vascular, respiratory, and metabolic disorders were documented in this selected "glaucoma-free" population at the VA facility.

Hypertensives were then categorized at the VA as patients with a systolic pressure over 180 or a diastolic over 100!

Considerations

R.L. Stamper, in 2007, considered whether *treated* systemic hypertension affects the progression of optic nerve damage in glaucoma suspects: "Systemic hypertension *treated with hypotensive medications* may be a risk factor for *increased progression* of optic nerve parameters in glaucoma suspects compared with age-matched normotensive subjects. Patients with systemic hypertension showed a statistically significant increase in cup area (cup-to-disk area ration and decrease in rim area, rim-to-disk area ratio and global Retinal Nerve Fiber Layer (RNFL) thickness ($p = 0.008$) with time…four years."

Blood Pressure and Glaucoma. A. Werne, A. Harris, D. Moore, I. Benzion, and B. Siesky (2008) question, "Are circadian variations in systemic blood pressure, ocular perfusion pressure, and ocular blood flow, risk factors for glaucoma?"

The **Hypertensive Diabetics** in the 1962–1967 VA study. Overall their IPOs were 13.5 in the first eye, 13.0 in the second eye, a consensual drop of 0.5 mm by the testing for two minutes of the first eye, with better aqueous outflow facilities than the alcoholic diabetics.

Alcoholic Diabetics had the least favorable IOP and lowest outflow facilities of any group.

Peptic-ulcer diatheses, acute or recurrent, on or off medication, these twenty-eight individuals had the expected initial ocular pressures, but significantly lower aqueous outflows.

Vagotomized patients. The thirty-nine patients who had undergone vagotomies and pyloroplasties within the prior ten years had significantly lower initial pressures and somewhat higher outflows than might be expected based on their ages. The vagotomized patients were somewhat younger than those in the medically treated group, yet they lacked consensual responses in most of the thirty-nine cases.

Inference: impaired parasympathetic input? While for the twenty-eight medically treated peptic cases, their consensual responses were average ~ 0.4.

"**Autonomic drugs in ophthalmology.** Some problems and promises." Directly and indirectly acting para-sympathomimetic drugs (Holland M.G., 1974).

Circadial Influences on IOP?

State-dependent and interactive variables may influence intraocular pressure changes during sleep. Recumbent posture, dark adaptation, and decreased sympathetic tonus are considered.

Findings

A.J. Sit (2008): "Recent researches indicate that intraocular pressure (IOP) does not decrease significantly during the nocturnal period…although the aqueous humor flow decreases by 50% or more at night."

A.J. Sit (2009): "IOP is highest at night and lower during the daytime, largely due to changes in body position, likely due to circadian variations in sympathetic nervous system activity."

J. Choi (2006): "Intraocular pressure depends on a nyctohemeral rhythm and in healthy subjects is higher at night than during the day, with a nocturnal peak value."

In glaucoma patients, however, the twenty-four-hour IOP rhythm was shown to be reversed, with values higher during the day (a midday peak in IOP) than during the night.

J. Choi (2009): "Decreased Nocturnal BP in 54 untreated Normal Tension Glaucoma (NTG) cases…the structural damage in eyes corresponded to the degree of reduction in their nocturnal mean arterial pressure."

A. Harris (2009): "Clinicians cannot currently visualize ocular blood flow directly…[yet] they can easily measure glaucoma patients' BP and IOP to calculate their ocular perfusion pressure and quantify the vascular changes."

However, estimates of gross blood flow levels or pulsatile flow do not necessarily indicate metabolic competence, for shunts and capillary closures may vitiate such conclusions, as is evident in the perfusion defects in diabetic retinopathy.

Clinical Application

The presentation in this EPA40 Series describes a noninvasive and clinically applicable means for assessing the ocular blood flow changes made in response to briefly imposed photic stresses or to therapeutic interventions.

With flicker at 40 Hz, a physiological purple entopic annulus manifests itself promptly in healthy eyes; it is delayed or absent in the impaired eyes. The critical state-dependency issues and a response indexing measure are addressed.

The genesis of neuropathy in open-angle "idiopathic glaucoma" is attributed to impaired ocular perfusion.

Part Two

Entopic Manifestations of the Semantic Software

The Dynamic Engrams (Motion-Memory) (Topic C7)

Quo Vadis?

"Motion-memory…an activity beyond perceptual processes."

Tulving, 1990

"For subsequent deployment in cognitive judgments…a motion memory system must…necessarily be maintained…with some degree of accuracy."

Blake, 1997

"Retrieval and use of relational memory depends critically on the hippocampus and occurs obligatorily, regardless of response requirements."

D.E. Hannula, 2007

Ida Pearce, M.D.

Abstract

Recursive engrams are the covert off-line substrates which support the conscious perceptions of dynamic online-scenes. Explicit evidence of recursive visual memory systems is obtained when viewing dynamic events, or following the idiocentric saccadic reading of texts.

Primary engrams reach conscious perception with eyes closed in room lighting being rendered visible only in a stochastic resonance attained in this umbral viewing mode. Engrams are generally imperceptible in the dark or in fields homotopic with online images, yet they remain accessible to cognition.

Dynamic Engrams are semantic images registered in parity with their objective sources, a distinction from the Motion After-Effects phenomena (MAE). The Recursive Engrams as witnessed are achromatic and of low resolution, yet replicate off-line those features observed online and within one context.

Regardless of whether the scene had been observed attentively or casually, its features remain in register, retaining form-with-motion in a unified display.

Whereas the scene itself had itself been scanned, the Dynamic Engram remains unscannable; its psychic projection shifts with head or eye motion in response to ongoing vestibular or oculomotor signals.

Engrams are explicit only in an "umbral view," with the eyes being closed but illuminated. These phenomenal visual engrams, witnessed off-line may persist for up to 120 seconds (Glenn A. Fry, 1969).

Present Purpose: To introduce unexploited means for subjective access to semantic recursive engrams. A literature search has not disclosed reports of these visually explicit entoptic phenomena, which are imperceptible in the dark or in fields homotopic with online images. These phenomenal visual engrams, witnessed off-line, may persist for up to 120 seconds (Glenn A. Fry, 1969).

Entoptic images are introspective experiences of endogenous patterns, generated either by innate hardwiring or by inculcate software.

Recursive motion engrams are clearly distinguishable from the chromatic flash photic after-images, from "imagery," and from the reversed MAE.

Evidence of Motion-Memory Engrams

In 1971, Buresová, Bures, and Rustová published papers entitled:

"*Acquisition and Retrieval of Visual Engrams*"

"*Consolidation of Visual Engrams*," and

"*Interocular and Interhemispheric Transfer of Visual Engrams in Callosotomized Rats*"

According to X. Huang et al., in 2008, "A serendipitous observation led to the study of V1 activity rebounds, which occur well after stimulus offset...and may provide new insight into the dynamics of early visual processing."

I myself first encountered the "Semantic Motion-Memory Engrams" on September 8, 2001, an experience which prompted these ongoing studies of mnemonic phenomena.

During a UCTV seismology lecture, a sequence of dynamic graphic images was being presented at 7:00 a.m. This TV program had featured two talking heads and four graphic segments in which each event happened to illustrate a different motion mode.

At the conclusion of each pictorial segment, my eyes were then lightly closed while listening to the speakers, and to my surprise the recursive image representations of each preceding episode shortly emerged to view. Each "illusory motile event" was witnessed for about ten seconds, but in the interest of viewing the next audibly announced upcoming graphic sequence, my eyes were reopened, thus the "after-images" were not left to expire, or to reiterate. Following each observed motile event, and following the eye closures, a motile engram shortly appeared, and an "illusory rerun" or replay was thus witnessed.

The TV program provided four graphic sequences:

1) Concentric centrifugal seismic surface waves were graphically superimposed on a map of the San Francisco area (oscillation/radiation). This map was stated to be replayed three times at triple real speed.
2) Graphics simulated a sinusoidal swaying motion of a tall structure.
3) Arial camera views of Bay Bridge showed sway and angulations at the two expansion joints.
4) Views from the air of traffic moving (linear translation) on the bridge, and with a changing perspective of the camera angle, optic flow (point of view) with parallax.

The sketch below and the following appended, unabridged comments were promptly inscribed:

Onset. Following each of some ten seconds of the visual priming during each graphic segment and after a latency delay of three to four seconds following eye closure, illusory rerun or replay was seen.

Morphology. There was apparently a faithful representation of the contours and topology of the depicted structures.

Rendering. This was achromatic, grainy, with only fair resolution. These motile semantic images appeared as monochrome or sepia, resembling old Daguerro-type photographs, though in a movie forma.

Motion. These appeared to be in real time, and the various dynamics duplicated my immediate recollection of the original visual experience of each of the four-event segments. The note then asked: Matching recollection...This recognition provided from what other mental repository? And how many repositories? Van der Velde et al. (2008) models, "Multiple interacting instantiations of neural dynamics as a highly modular multi-level C++ framework."

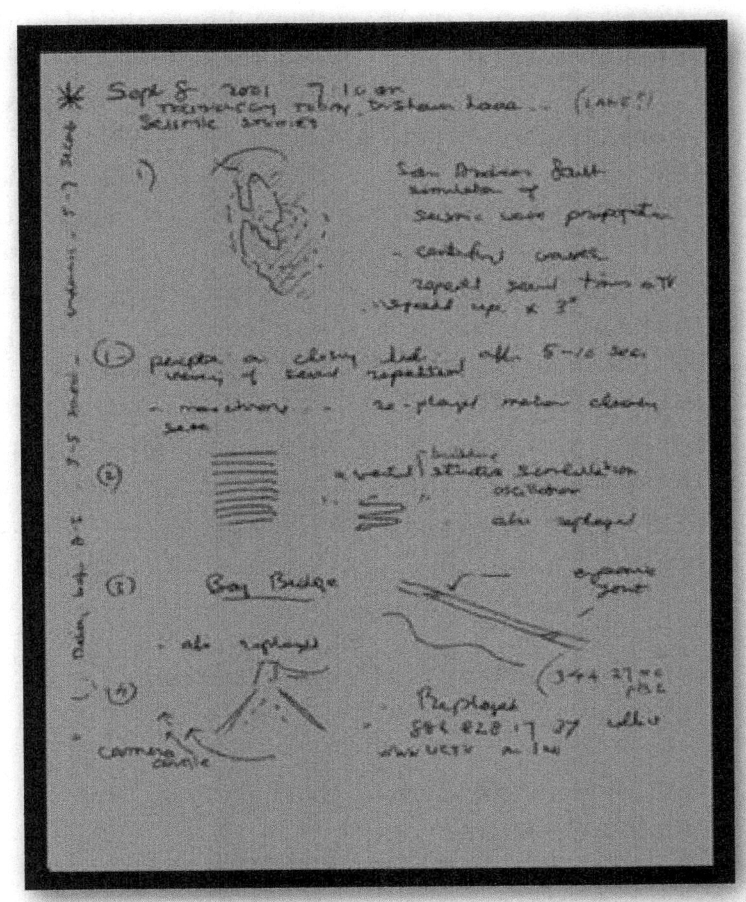

San Andreas Fault Simulation of seismic wave propagation, centrifugal waves, repeated several times on TV "sped up x 3." Perception on closing lids after 5-10 second viewing of several repetitions. Monochrome replayed motions already seen. Original notes by Ida Pearce, M.D. 2001.

September 18, 2001. This second engram event was experienced and was documented more briefly: a persistent replay of TV tickertape with motion...depends on the ambient light through lids...sustained for one to two minutes...with positive image...none seen in the dark.

The above events were unprecedented experiences of a phenomenal visual motion-memory system. Those initial findings have been substantiated to date, 2008, with confirmation of temporal parameters and state adaptations, notably the umbral status with the ambient light through closed lids.

According to Turk-Browne et al. (2005), "It requires attention to select the relevant population of stimuli, but the resulting learning then occurs without intent or awareness." In the simple protocols outlined below, linear translation was selected for its convenience and the availability on commercial TV of the empirically designed, nonergodic horizontal presentations, which included banners, news clips or tickertape, and a variety of unstandardized ascending credits.

Fredericksen and Hess, in 1997, found, "The spatial properties of motion sensitive cells (human) match to the statistical properties of 'natural scenes'; velocity peaks ranging from 0.25 to 5.66 c/degrees."

Dynamic Motion-Memory Engrams

Protocols:

One

1. Induction and Longevity of the Engrams
2. Outcomes from Intervention
3. Subsequent Vestibular Modulation
4. Explicit Replay of Spreading Textures
5. Engram Perceptions with Subjacent Online View
6. With Asymmetric Ocular Input (Interocular Transfer?)
7. Optic Flow and the Motion Engrams

Two
Natural Scenes, with Autonomous Mnemonic Registration
"Ecologically Valid Stimuli"
Dynamic Visual Engrams (DVE) vis-à-vis the Motion After-Effect

Three
Dynamic Visual Engrams vis-à-vis the Motion After-Effect Phenomena

Part One of Protocols

Protocol One
Induction and Longevity of the Dynamic Visual Engrams

Mesopic adaptation and mesopic ambient light are generally favorable for the generating and witnessing of robust engrams; "learning then occurs without intent or awareness" (Turk-Browne et al., 2005).

In the simple protocols below, linear translation was selected for its convenience and the availability on commercial TV of the "empirically designed" nonergodic horizontal news clips or tickertape.

Step 1: The timed exposure selected here is ten seconds, ample for an engram capture of rolling credits or of tickertape. With any commercial TV screen set at a convenient distance, either a fixed gaze or randomized viewing strategies may be employed. During normal saccadic viewing, a fixation sampling rate may be around one to three per second; this is considered a low sampling rate for the detection of motion. According to Mermillod et al. (2005), "A faster integration of low-spatial frequency starts in early retinal processes, as compared with high-spatial frequency."

Step 2: At ten to fourteen seconds, the eyes are now closed in room lighting. Only the salient reset after-images of any diffuse, luminous areas are at first perceptible, but their contours fade over a couple seconds.

Step 3: At sixteen seconds there is now the perception of a positive engram. Seen only with the closed illuminated eyes, these motile images emerge as achromatic and granular. In this umbral view, the engram may reiterate for up to 120 seconds.

Protocol Two
Outcomes from Intervention

This second sequence demonstrates the outcomes from interventions applied to an already established, visible engram, as identified by Step 3 of Protocol One.

Step 4: When complete occlusion is applied for several seconds, the only visible pattern remaining in the dark is now the intrinsic rosette template, q.v., and a black background with luminous blue cluster oscillations. The engram pattern itself is imperceptible in the dark.

Step 5: Now, again with the illuminated closed lids, the engram is restored to view.

Step 6: But next, with momentary full exposure to room light or to an articulated scene, the engram is erased.

Step 7: On restoring illumination to closed lids, the engram now is nonrecoverable, irrevocably erased from view.

On Disruption:

Peacock Miller (1982): “Eidetic imagery is critically dependent on level of illumination, and its contents are easily disrupted by after-coming visual stimuli.”

Verstraten et al. (2007): “Disruption of implicit perceptual memory by intervening neutral stimuli is reported.”

On Visible Longevity:

Without an intervening, disruptive visual input, a recycling engram expires only when its “working memory life” is exhausted, apparently at a maximum of two minutes. The cycling is terminated earlier by a reordering of the veridical visual environment, such as a novel visual event or increase in ambient light. At Step 3, darkness evidently did not signal for cancelation of looping, for although an established engram is not visible in complete darkness, it can be restored to view at Step 5, if initiated within the time limits of its potential two-minute longevity. As in a computer, there may be one terminator within the nested loop (inherent time limited) and other outside terminators, as Step 5, here a novel or updated visual event.

On the Delayed Conscious Perception of the Engram:

Although there is perceptual impenetrance in the four-second "delay phase," the engram was already deployed for predictive judgments and precipitous responses. According to Kleinschmit et al. (2002), "In observers exposed to gradual contrast changes of stimuli...brain activity was recorded with functional neuroimaging stimuli prior to its sensory perception."

The engrams of dynamic scenes seen in achromatic coarse rendering are believed to be channeled for prompt autonomous deployment. According to Goodale et al. (2005), "Many action tasks have strict temporal constraints, which can only be met if the visual information is relayed directly to the motor system without first passing through a conscious decision-making process."

Harter and Kozma, in 2005, introduced "systems capable of replicating the important principles of a periodic/chaotic neurodynamic while being fast enough for use in real-time autonomous agent applications." The delayed visible recursions reported in the umbral viewing mode are adventitious, inadvertent, and only incidentally perceptible to conscious awareness while en route to other cerebral addresses (as read-only?)

On Engram Content and Spatial Parity:

Engrams retain the precise spatial placements imposed during induction by the contemporary inputs from online sensorimotor heteromodal inputs. (See Protocol Three.)

On the Coarse Engram Rendering:

According to Mermillod et al. (2005), "The human perceptual system seems to be driven by a coarse-to-fine integration of visual information; results have shown a faster integration of low-spatial frequency lossy compared with high-spatial frequency information, starting at early retinal processes." Principles at coarse scale organize the cortical spatial arrangements.

Protocol Three
Vestibular Modulations and Off-line Perceptions

- On TV, vertically rolling credits are commonly presented. These may either be read or observed casually for ten seconds while the engram (as a persistent dynamic image) is being generated.
- Then with closed, lit eyes, the up-going lines, now as an engram text, can be consciously monitored.
- The apparent angular momentum of the ascending credit lines may then be matched by a deliberate, smooth pursuit, made with head motion only.
- While elevating the chin to track the motion through some seventy vertical degrees, these up-going lines thus appear as if frozen in space, linked with the idiothetic heading.
- At the azimuth, the tracked images may still appear to remain "immobile" for a further four seconds while the maximally elevated head posture is steadily maintained.
- After this four-second delay, the up-rolling motion of the lines of text resumes at the original pace while the elevated head remains immobilized. Clearly the oculomotor-vestibular input operated during an engram acquisition, and a vestibular effect, was also demonstrable after the initiating "priming" event had concluded, as could be detected thereafter as an "updating" of a visuospatial engram.

Engrams retain the precise spatial placements imposed during induction by the contemporary inputs from online sensorimotor heteromodal inputs, including the vestibular inputs that operate both during the initiating "priming event" and demonstrable upon the recursive engram display. Vestibular inputs can also modify the Helmholtz Traveling Waves (HTW). These slow wave fronts traverse an angular subtense of approximately 180 degrees in about four seconds, but during a brief four-second smooth pursuit made by head motion, they can exhibit the interactive vestibular phenomenon of a motional hiatus. The "long wavelength" (greater than 15 cm) traveling waves propagate from occipital to prefrontal electrodes" (Srinivasan, 2006)—putative classic HTW.

Vestibular Phantom Grids. Under passive rotations with closed eyes, it appears that changes in the direction of angular momentum may enable the perception of a "laggard phantom grid." This unreported phenomenon is attributable to an asynchrony between the slower retinal and the faster labyrinthine-sourced transmissions, which accounts for the noted temporal lag.

Postural Changes. A Foveal After-Image (FAI) is imprinted in primary horizontal gaze, then after three to five seconds in motionless down-gaze, this FAI image appears to drift slowly upward at a rate of some ten seconds over the ninety degrees. If, however, a smooth pursuit is made by slowly elevating globe and chin, this "target" FAE can be tracked upward; it tends to stabilize as horizontal centering is attained. If instead the proclivity to pursue the FAI by smooth pursuit is denied, a linear chain of luminous nodes may now materialize below the "captive FAI."

Alternatively, with a ten-second horizontal FAI installation, and then by immediately elevating the gaze to ensconce FAI at an up-gaze starting point, the scenario is reversed: there is a down-drift of the after-image toward horizontality. These slow drifts presumably are generated by otolithic somato-centric responses, signaling both oculomotor and cervical and spinal musculature.

By changes in vestibular angular momentum, the apparent dynamics of an established engram may be modified; the engram's dynamics remain unaffected by varied oculomotor activity, but the engram may be disrupted.

Loose, Probst, 2001. "Angular velocity, not acceleration of self motion, mediates vestibular-visual interaction."

Bertin, Berthoz, 2004. "Visuo-vestibular interaction in the reconstruction of travelled trajectories. We show for the first time that a vestibular stimulus of short duration can influence the perception of a much longer-lasting visual stimulus."

Trappenberg et al., 2005. With single neuron recordings found with primates that 'hippocampal spatial-view cells…not only maintained their spatial firing in the absence of visual input…' (motion memory sustained!)…but could also be updated in the dark by idiothetic input."

Wei et al., 2006. "Visuo-spatial updating uses vestibular information. Intact labyrinthine signals are functionally useful for proper visuo-spatial memory updating during passive head and body movements."

Fetsch et al., 2007. As a function of the relative strength and spatial congruency of visual and vestibular tuning, the reference frames in the combined condition varied as a function of the relative strength and spatial congruency of these inputs. Tuning for optic flow was predominantly eye-centered, whereas tuning for inertial motion was intermediate but closer to head-centered. (See Protocol Six.)

Smooth Pursuits

A vessel at sea is observed on a TV screen; the resultant engram is then viewed. Either the moving vessel or the receding wake may be selected as the target for smooth pursuit in the engram.

September 13, 2005. By postural corporeal smooth pursuit, I observe flocks of pigeons wheeling across some 180 degrees. A visual engram replay may be deliberately accompanied by a replication of the combined postural and head motions, a vestibulo-motor memory. Such behavioral responses may be videotaped. De'Sperati in 2005 recorded by infrared oculography the eye movements during the "mental extrapolation of (memorized) saccadic motion." The subjects were primed by targets oscillating sinusoidally by plus or minus five degrees on the horizontal plane, at frequencies between 0.15 and 0.5 Hz.

DeLucia, 2006. "Mechanisms that underlie boundary extension and representational momentum…contribute to the integration of successive views of a scene while this is changing."

Ruiz-Ruiz, Martinez-Trujillo, 2008. "Human Updating of Visual Motion Direction during Head Rotations."

IXP, 2008. The idiothetic updating of motion engrams is demonstrable.

Protocol Four
Explicit Replay of Spreading Textures

After idly gazing at graph paper, at a shag rug, a sleeping dog, a page of print, or a complex dynamic scene, the unique textural character of each field may then be distinguished with closed, illuminated eyes. After a couple seconds' delay, these textures emerge as positive-featured, achromatic images of low resolution across the whole field.

Theories of Textural Discrimination and of Spreading. Tyler (2004): "A roving local sampling window…allows the visual system to derive from any particular texture image an estimate of the ensemble statistics over the window…without the need to present multiple samples for evaluation…single neurons in V1 can signal the presence of higher-order spatial correlations in visual textures…This places a computational mechanism, which may be essential for form vision at the earliest stage of cortical processing."

Dynamic Texture Spreading. Wollschlager and Faul (2006): "…probing the mechanisms of surface interpolation…indicates that information fragments are integrated over a time window of about 100 to 180 ms to form a complete surface representation."

Directional Harmonic Theory. Lehar, 2003, described "a computational Gestalt model to account for illusory contour and vertex formation."

Integration of Sequential and Related Texture Patterns. Compatible recursive engrams exhibit a perceptual transparency/percolation.

September 23, 2001. Observing TV screen for several minutes, fixation ranged over the talking heads and over the text. By viewing tickertape with three horizontal lines, ciphers streaming to the left, the motile Engram A was established and verified as a persistent image seen through a red lid—continuous movement of an illegible after-image.

Next, with the head tilted forty-five degrees to the left, a second motile Engram B was acquired. Then, with eyes closed, both sets of lines were represented simultaneously. Note then reads on reversing head position...the archived surviving after-image crisscrosses the recent imprint. The final sketch indicates horizontality restored to the original engram and tilt of the overlying second "imprint" as seen in the composite umbral view. Fluctuating, patchy integrations were seen between successive arrays represented in the two sustained mnemonic visual images.

According to Wandell et al. (2005),"Exact site percolation thresholds using a site-to-bond transformation, and the star-triangle transformation."

December 22, 2003. It was noted and then documented that after sketching or writing in ink on a page of standard lightly ruled one-quarter-inch graph paper, the overall orthogonal texture geometry (grid) persists in memory seen with closed, lit eyes. Curiously, the darker ink inscriptions on the graph paper were rendered coarsely and illegibly, whereas the overall grid lines remained clear and crisp. (Note: A fixed gaze is not required to establish these engrams.)

April 26, 2007. The above phenomenon in 2003 was recently duplicated and subjected to further scrutiny. An array (the standard graph paper itself) is presented for ten seconds for each of two orientations in sequence. Each presentation retains a mnemonic visual image (engram) to be verified promptly during the sequential umbral views, thus:

1) *The First Array.* The overall orthogonal texture geometry (grid) persists in memory seen with closed, lit eyes; the darker ink inscriptions on the graph paper again appear coarse and illegible, while the overall grid remains clear and crisp. The precise fine grid geometry is also clearly perceptible against a brightly lit screen, being perceptible then only with the imposed flicker. Repeating the 2003 experiences of the First Array, its texture engram is then verified, having been set up as a foil for this next event.
2) *The Second Array.* This is now the same target, reoriented; the graph paper being rotated through forty-five degrees. On eye closures, the engram of the first orientation persists, incorporated with the second engram. Thus, in umbral view, and also in the flicker-viewing condition, these two engrams compete, seen as fluctuating patches, with the squared texture areas either in rivalry or mutually engaged as integrated patches of octagons.
3) *Rescaling the Second Array.* Easily accomplished by doubling or halving the viewing distance from the target grid, this ploy also demonstrates fluctuant incorporations.
4) *Moiré Patterns.* These were seen to develop spontaneously while witnessing the engrams (two plus three above). These moiré engram patterns were also susceptible to modulation by imposing flicker onscreen.
5) Online moiré patterns, which develop during an online viewing of an orthogonal metal-mesh window screen,

typically emerge with periodic magnifications of about 3.3 scales, and at forty-five-degree rotations, indicating intrinsic processing architecture.

6) Transparent incongruent arrays can also be incorporated as engrams. A one-quarter-inch piece of graph paper is placed over the monitor screen, which displays a large regular array of desktop icons, or other "texture" display. These incongruent textures can then be maintained as an engram, demonstrable in umbral view and also with flicker.

Deviation Detectors. Primitive sensor mechanisms have evolved to discern deviations arising in natural environmental textures. Human skills in pattern minutiae recognition have culminated in sophisticated discriminations, enabling the reading of texts. (This parallel mnemonic system is apparently dedicated solely to graphic-lexical inputs, proceeding "from texture to text.")

Protocol Five
Engram Perceptions with Subjacent Online View with Half-Lowered Lids

With vertical hemi-field segregation, can exogenic and endogenous images simultaneously enter perception?

Berger et al. (2005): Competition between Endogenous and Exogenous Orienting of Visual Attention.

December 19, 2001. Perceptions from vertically separate fields. In up-gaze, with lowered lids occluding only the upper field, the images of the real objects in the lower field of view may be projected normally into external space and constitute the "object-based frame of reference" for the observer. Meanwhile, an acquired engram may still persist, vignetted behind the half-lowered upper lid; shielded in this upper field it is retained in the umbral view. The two adjacent images show no transparency or overlap, and the engram preserves its own umbral domain sector within the cognitive field. This engram image may become disrupted by eye motion, but not by smooth head motion.

Explicit perceptions of the environment and perceptions of recursive engrams are state dependent, and are mutually exclusive.

With heterotopic fields, an engram is perceptible with an online subjacent image. The umbral status is equivalent to the introduction of noise—stochastic resonance—"the counterintuitive phenomenon in which noise enhances detection of sub-threshold stimuli" (Perez et al., 2007). Yet an off-line engram can persist if sheltered in a secluded umbral view (i.e., if not in the same phase-space).

Sdoia et al., in 2004, investigated "the relation between visual hemi-fields and spatial frames of reference, according to the idea that multiple representations of 3-D space exist."

Results from two experiments clearly show that an upper-visual hemifield advantage only arises when allocentric spatial judgments are required in order to perform a location task, whereas a lower-visual hemifield advantage arises when egocentric spatial judgments are required.

Protocol Six

With Asymmetric Ocular Input (Inter-Ocular Transfer?)

Apparent Inter-Ocular Transfer. This phenomenon emerges following articulated input to one eye, with low-luminance input to its fellow, and relates to the role of the stochastic noise generated in the fellow eye.

"Acquisition and retrieval of visual engrams" and "Consolidation of visual engrams."

O. Buresová, J. Bures, and M. Rustová (1971): "Interocular and interhemispheric transfer of visual engrams in callosotomized rats."

J.E. Raymond (1993): Complete interocular transfer of motion adaptation, effects on motion coherence thresholds.

B. Timney et al. (1996): "We conclude that the type of occlusion used for measuring IOT is important only when visible contours in the non-adapting eye contribute to the adapting process."

"IOT does not appear to strongly depend on conventional binocularity of neurons."

C.M. Howarth (2008): "Interocular Transfer of Adaptation in the Primary Visual Cortex."

R. Tao, M.J. Lankheet, W.A. van de Grind, and R.J. van Wezel (2003): "It is well established that motion aftereffects (MAEs) can show interocular transfer (IOT); that is, motion adaptation in one eye can 'give' a MAE in the other eye."

The stochastic noise transmitted from the "umbral eye" lowers the threshold to conscious perception of the cortical engram. This engram image is thus referenced to but is not retrograde in the other eye.

Default views here present the following perceptions:

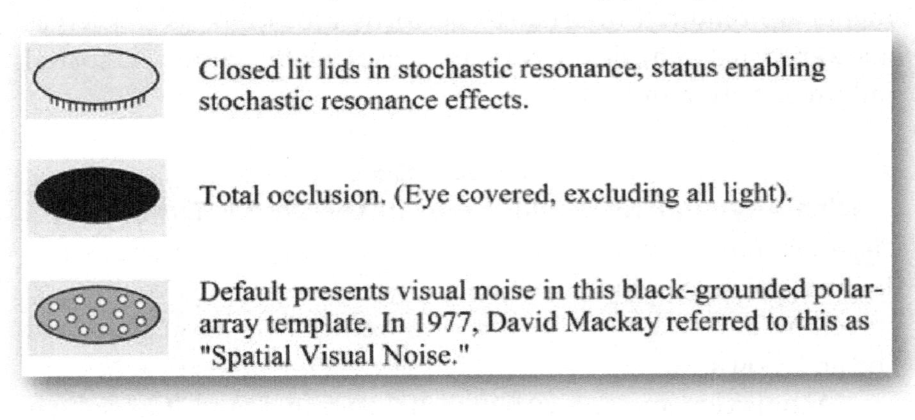

Engram perceptions in Conditions III, V, and VI.

Protocol Seven

Inter-Ocular Transfer with Asymmetric Ocular Inputs

The stochastic noise transmitted from the "umbral eye" lowers the threshold to conscious perception of the cortical engram; this engram image is thus referenced to but is not retrograde in the other eye.

Lehmkuhle and Fox, in 1976, had reported that interocular transfer (IOT) of a translational- motion after-effect was greater if the nonadapted eye viewed an equiluminant field (Step V) than if it viewed a dark field (Step VI).

The engram motion-memory replicates the vector of the stimulus, not in the reverse direction as perpetuated in the widely studied MAE.

The Following Steps Occur at Approximate 5-Second Intervals

Processes Evoking Visual Perceptions Perceptions	Step	Left Eye Status	Visual Perception	Right Eye Status
Default in both eyes	I			
Motile Input to right eye only	II			
Engram Perception views by right eye only	III			
DEFAULT in both	IV			
Engram now apparent to left eye only	V			
Engram apparent now to right eye only.	VI			
WHITE SCREEN to right and left eyes Engram is erased	VII			
DEFAULT	VIII			
RIGHT and left eyes	XI			

Steps:

I. This protocol begins with both eyes in the umbral status (I).

II. The *engram* is then induced monocularly, via an open right eye (II).

III. Verified briefly in umbral status, while the fellow left eye remains fully occluded throughout this process (III).

IV. On full occlusion of both eyes (IV), the engram perception is briefly absent.

V. When umbral status is then promptly conferred upon the left eye, the engram is sustained. It appears referable only to this left eye; while the right adapted eye remains fully occluded and signals only its own black-grounded polar-array template. David Mackay, in 1977, referred to this as "Spatial Visual Noise." The umbral status is equivalent to the introduction or inherent stochastic noise.

VI. The engram perception is obviated with full occlusion of both eyes, as in IV.

VII. This engram is, however, visually recoverable in umbral views, if sought within its 120-second perceptual-longevity limit (VI). Refer back to Protocol One, Step 3.

VIII. Exposing both eyes to the light erases the perception of the engram (VII). Refer back to Protocol Two, Disruption of an Explicit Engram by Intervening Stimuli.

IX. Neither a reverting to umbral status of both eyes (VIII) nor a total occlusion (IX) restores perception of the engram.

The umbral status is equivalent to the introduction of noise, "a counterintuitive phenomenon in which noise enhances detection of sub-threshold stimuli" (Perez C.A. et al., 2007).

According to J. Ding and G. Sperling (2006), "Each eye may exert gain control on the other eye's signals in proportion to the contrast energy of its own input, and additionally exerts gain-control on the other eye's gain-control." Stalemate?

"The global percepts with closed lids and dichoptic diffuse light: Adaptive Transients and Filamentary Flashes and Stable Dots," as noted in 2000.

Conclusion: The stochastic noise transmitted from the "umbral eye" lowers the threshold to conscious perception of the cortical engram; this image is referenced to but is not "in the other eye." The cortical engram is registered with the visual noise arising from the "nonadapted" noninstructed eye, the default eye.

According to Tao et al. (2003), "It is well established that motion-after-effects (MAE) can show inter-ocular transfer (IOT), that is, motion adaptation in one eye can give a MAE in the other eye."

Therefore, stochastic noise transmitted from the umbral eye lowers the threshold to conscious perception of the cortical engram, thus referenced to, but not **in** the other eye.

Protocol Eight
Optic Flow and the Motion Engrams
Streaming and Smearing of Texture

Motion Streaks Provide a Spatial Code for Motion Direction: According to Geisler (1999), "Spatial signals for motion direction exist in the human visual system for feature speeds above about 1 feature width per 100 ms."

June 27, 2006: Solivitur ambulando? On walking along a coarse gravel path, its texture is equivalent to a grating of contrast approximately 30 percent and intervals exceeding an inch. This surface is viewed with a fixed gaze on an outstretched finger on vertical down-gaze; while keeping a smooth pace of about one meter/sec for fifteen seconds, the surface is seen as down-streaming.

On walking backward with a normal front-gaze, the optic flow appears convergent. On walking backward in down-gaze, the surface of the path appears to be up-streaming, as then appears in the engram. Then, in smooth pursuit of this engram by chin elevation with closed eyes, the apparent motion ceases for some seconds then resumes its upward motion (as was seen in Protocol Three.)

Circling beneath a leafy tree—Wednesday, June 21, 2006—with ten seconds of free viewing and with confluent head motions made from a down to upward gaze. On eye closure, the engram images seem sparse and noisy with respect to local motions of these large areas. With large receptive fields for optic flow detection in humans, "sensitivity for radial, circular and translational motion is increased with stimulus area at a rate predicted by an 'ideal integrator' in humans" (Burr et al., 1998).

According to Tyler, in 2004, "A theory of texture discrimination incorporating a roving local sampling window allows the visual system to derive an estimate of the ensemble statistics over the window from any particular texture image, without the need to present multiple samples for evaluation."

According to Barron and Srinivasan, in 2005, "In honeybees, manipulation of the visual texture (in wind tunnels) revealed that headwind is compensated almost fully even when the optic flow cues are very sparse and subtle demonstrating the robustness of this visual flight control system."

Part Two of Protocols

Natural Scenes with Autonomous Mnemonic Registration

Ecologically Valid Stimuli

Tulving and Schacter, in 1990, in their Persistent Representation System, note motion-memory as "an activity beyond perceptual processes," which processes they suggest "evolved to perform only ecologically valid computations."

According to Dan et al. (2005), "The use of natural stimuli is vital for our understanding of sensory processing."

The following anecdotal examples illustrate the broad range of confluent motions and dynamic events that may be captured autonomously as engrams and seen then as autonomous, reiterating recursions.

1) **Vortices with Countable Rotations.** The particles or bubbles on stirred coffee can act as markers. The colorful meteorological graphics on weather TV may also replicate well as achromatic engrams.
2) **Aquarium Displays.** With relatively dark-adapted observers, aquatic dynamics are faithfully rendered in umbral viewing. These venues offer ideal conditions for engram capture, presenting confluent streamlined flows, particulate diffusions of fine luminous particles in high contrast, and a wide range of organic biological activities.
3) **Conflagrations.** A refinery stack ablaze: As this engram recycles, the columnar uprising, the vertical component, may outlast the perceptions of the peripheral lateral turbulences, perceptually progressing to a motion-averaging effect. According to Burr et al. (1998), "Local speeds are averaged, independent of direction, to derive a global speed estimate...evidence for local averaging within, but not across two speed ranges. This finding established the existence of at least two independent speed-tuned systems in the range of speeds tested."
4) **Counter-flowing?** Castet and Morgan, 1996, state "plaid motions initiate motor responses made by summing of two orthogonal moving gratings." Viewing crisscross

palm fronds in asymmetric motion, the perception is *not* one of summation. Can plaid motions exist in nature? See Bloomberg's streaming text: three channels, two speeds, maintained in engram; see TV. Linear streams are also distinguished in engrams after the viewing of vehicular traffic patterns. With transparent spheres, the apparent counterflow of the nearer and further hemisphere of rotating Buckey balls, distinguished in engrams, a counter-streaming is perceived unsummed. The chiralities in axial views reverse when observed in up or in downward gaze at the pole of a rotating item.

Angular Velocity: According to Caplovitz et al. (2006), "The perceived angular velocity of an ellipse undergoing a constant rate of rotation will vary as its aspect ratio is changed." The view of transparent spheres is as of rotating ellipses.

Taraxacum: Dandelion seed-heads are 2.5 cm in diameter; the seeds are hexagonally distributed over the sphere and dispersed at 3–4 mm intervals. With each rotation executed in four seconds, the dark seed-dots are discerned to move coherently with angular velocities dependent upon their tangent to the proximate views of elliptical sectors, the poles representing singularities, axial pivot points.

5) **Screen Saver Designs**. Small, high-contrast particulates in loomings or zoomings simulating optic flows in 3-D are accurately portrayed thereafter in engram; mechanical or pendular oscillations do not prove effective. (Not ecologically valid computations?)
6) **Shifting Landscapes; Optic Flows and Panoramic Views.** According to Olveczky et al. (2003), "Cells selective for differential motion can rapidly flag moving objects, and even segregate multiple moving objects."

7) **Vehicular Travel.** July 17, 2005. For a passenger, travel through spectacularly diverse terrain provides semantic and dynamic complexity with layered and intersecting contours, distant horizons, and speed-blurred images as lines of fence posts approach with increasing angular velocities. As the varied contours of natural landscapes and structures occlude or unfold in parallax, sequences of the serial vistas of rolling intersecting landscapes are observed through sunglasses.

   Then, with closed, lit lids, each dynamic scene can replay as an engram, with high fidelity to the apparent relative motions. Blake et al., in 2003, offer a "counter-intuitive finding…that high contrast detracts from motion perception." The use of the dark sunglasses evidently contributed to the above impressively integrated effect. The larger the receptive field of a sensor, the better for motion detection (and also for the Kanizsa effect, which is enhanced at low luminances.)

   One mode "produces invariant representations of the motion flow fields produced by global in-plane motion of an object; above in-plane rotational motion, looming versus receding of the object" (Rolls and Deco, 2004). These dynamics are present in stereoscopic online experiences, one to seven above, and are witnessed in parity in the off-line, recursive, achromatic engrams.

8) **Seeing Biological Motion.** According to Neri, Morrone, and Burr (1998), "Information is summed over extended temporal intervals of up to three seconds by mechanisms that analyze biological motion and adapt to the nature of the stimulus."

   Primitive survival required the sensing of local scenes and activities and retention/persistence of these data. Not surprisingly, "the spatial properties of motion sensitive cells (human) match to the statistical properties of natural scenes…velocity peaks…ranging from 0.25 to 5.66 c/deg" (Brady, Bex, and Fredericksen, 1997). Hyvarinen and Hoyer, in 2001, describe "a highly parsimonious model whereby the properties of the visual cortex are adapted to the characteristics of the natural input."

Motion After-Effects (MAE) and Pictorial Memories

Much attention has been devoted over the years to the subjective yet noninformative reset phenomena of the illusory MAE. It is remarkable that available literature offers so few indications of the subjective, semantic, cogent recursive **engrams**. Apparently these recursive engrams were experienced and well described in 1982 by Miller and Peacock, who then wrote, "Eidetic imagery appears to be a long-lasting, percept-like experience which varies considerably in clarity and definition; its duration is critically dependent on level of illumination and its contents are easily disrupted by after-coming visual stimuli."

Replays of the phenomenal mechanical MAE have been verified by fMRI (Bouman and van de Grind, 2004). The subjective off-line informational engrams await similar objective validation.

Motion-After-Effects, vis-à-vis the Recursive Dynamic Visual Engrams

According to S.N. Watamaniuk and S.J. Heinen (2007), "Adaptation to motion produces a motion-after-effect…where illusory, oppositely-directed motion is perceived…when viewing a stationary image."

The Recursive Dynamic Engram. These images are not sustained to perception when viewing areas of the cognitive field are occupied by exogenic online images.

Motion After-Effects. MAE are not seen in the dark, but are revealed by prompt confrontation with a "static or dynamic noise pattern" and only sustained to recognition in areas of an articulated field occupied by exogenic veridical online images. (Engrams are also invisible in the dark.) MAE reportedly reverses the prior trajectories, angular velocities and streaming of linear, radial, or spiral motions of the earlier motile inputs. The MAE reported in 1911 by Wohlgemuth noted its extended longevity in darkness. According to van de Grind et al. (2004), "Static and dynamic testing has confirmed this and other well-known MAE effects, with explicit longevities for up to 30 seconds, and longevity of 40 seconds with fMRI."

Recursive Dynamic Visual Engrams. These dynamics differ significantly from the MAE dynamics. Unlike the MAE choreographies, only the original trajectories appear in the engrams. Whereas the MAE initiation process has been considered an "adaptation," an appropriate term for the engrams is a "capture" or "acquisition." These captured primary engrams are covert images in the sense that they become explicitly visible only with closed, illuminated eyes after a four-second delay, and then they may iterate with visibly declining amplitude for up to 120 seconds.

These continuum fields with confluent motion and embedded coherent structures are replicated as forward-going, achromatic, and recursive images. According to X. Huange et al. (2008), "In addition to providing a possible explanation and neural correlate of a visual aftereffect...rebounding activity may provide new insight into the dynamics of early visual processing."

Verstraten and Ashida, in 2005, stated that "the visual system dynamically calibrates its internal bias...using a recent percept... this form of positive bias, or priming, is created in an automatic fashion." The autonomous acquisition of semantic engrams is considered fundamentally a banking of information in a short- or long-term deposit, rather than a "priming" process.

Engram images are not sustained to perception in areas of the cognitive field occupied by exogenic veridical online mages. A visual field, viewed with half-closed eyelids, enables perception of the external views in the lower field, while the motion-memory packages persist in the upper field.

These engram visual manifestations, witnessed subjectively and logged since 2001, have proven replicable and consistent with current theories on human visual memory acquisition-cum-learning. Memory-and-navigation systems are reported in many other animal species, including primates, rodents, birds, insects, and cephalopods. The semantic engrams in man might now also be validated by fMRI. Meanwhile, the opponency reset MAE phenomena "remains questionably of fundamental significance... and [has] enjoyed considerable and perhaps unwarranted attention" (Raymond, 1993).

Effective Confluent Inputs

Engrams can be generated during casual and brief viewings of dynamic scenes such as rolling credits or tickertape on a TV screen, watching the flow of traffic, watching a marching band, or stirring coffee. After viewing such a scene for three seconds, or for up to thirty-second periods, those features are then recognizable in the persistent motile images. Multiple motions are replicated with dynamic fidelity, though rendered achromatic.

A robust engram as brief as five seconds may then recycle with amplitude attenuations for up to 120 seconds to reach a "fuzzy endpoint." The stochastic noise transmitted from the "umbral eyes" lowers the threshold for conscious perception of cortical engrams. The engrams' longevities apparently exceed those reported of the MAE.

Oscillatory Inputs

Wohlgemuth, in 1911, stated "short-period oscillatory motions do not excite MAE." Confluent, natural motions favor initiation of the engram.

Mechanical oscillations are not replicated as engrams; such inputs to this system may cancel out, due to low sampling rates and to motion averaging. According to Lappin et al. (2002), "Motion produces visually coherent changes in image structure, but stationary contrast oscillation does not."

Recursive Lexical Engrams (Topic B5)

Reading is an Engram-Dependent Skill!

Present Objective: To offer hitherto unexploited means for perceptual access to "reactivated" recursive mnemonic images: the engrams. These explicit semantic engrams parallel the implicit **PRS** of Tulving and Schacter, 1990, defining two Perceptual Representational Systems with "Highly Idiosyncratic Spatial or Verbal Specificity." Following the viewing of fluid motions and dynamic events, or following the saccadic reading of texts, there is subjective evidence of these discrete mnemonic visual memory systems.

Cornelissen, in 1998, had described "Coherent Motion Detection" and "Letter-position Encoding."

P. Brugger, in 2007, noted that "Linguistic and spatial cognition are more tightly interwoven than is currently assumed."

Material: Standard texts in Times New Roman twelve points as viewed at 75 percent.

Methodology: Routine saccadic reading of text from varied distances.

FINDINGS:

Engrams: Autonomous memory-traces of a congruent event or of one page of text. Recursive Engrams prove accessible for autonomous processing and consolidations.

Textures: Extended fields include textiles, wallpapers, and microscopic histology. The print on this present page viewed in dim light might appear as texture.

Textons: According to B. Julesz (1981), "The term visual texton designates the perceptually-separable elements as viewed in any textured patterned field." Dimensions of textons vary within one field of any arbitrary scene and in a printed text.

Lexors: Autonomous cellular agents. These sensors, the texton-detectors, operate in synchronized temporal circuits, which are also spatially concentric.

ACME: "Archived, Compacted Multiple Engrams."

DATES OF MY ORIGINAL FINDINGS:

The Recursive Lexical Engrams - July 22, 2000
Autonomous Lexors - October 16, 2000
Engrams Triggered by Flicker - February 13, 2001
Spatial Motion-Memory, Dynamic Engrams - September 8, 2001
Lexical Archives with Shifting Engram Spreadsheets - 2003–2008

Theses I–X
The Lexical Engrams:
"Recurrent Dynamics in Excitable Media"

Thesis I: Semantic visual engrams are autonomously imprinted in the brain and may be witnessed as explicit in *off*-line verbatim recursions.

Autonomous Processing in the Default States

F. Esposito (2009): "The default-mode functional connectivity of the brain correlates with working-memory performances."

Thesis II: The Motor Vectors and Trajectories

The online versus the off-line sequential visual mechanics during saccadic *online* readings.

The neuromechanics of saccadic rightward motion by the oculomotor saccades, first moving rightward across the stationary text, are then directed leftward by a single oculomotor saccade, thus returning back to the next line of the stationary printed text.

During the *non*saccadic *off-line lexication* process:

When witnessed in Condition III, the *off-line* lexication of an embedded engram text, the letters and lines shift steadily leftward with **no** regressions. This flow enables scrutiny by the stationary cortical lexors.

Cyclopean Functions: See Schira et al., 2009, "...alignments with the polar-axial view."

Meanwhile, with fixed gaze, the engram text itself slowly rises vertically (as if printed on an ascending, leftward-rotating cylinder, as was first noted in 2000).

N. Haydn (2005): "We analyze a skew integrable map defined on a cylinder that models a shear flow."

Thesis III: Lexication *off*-line is *accelerated*, exceeding 1,000 WPM. Reading processes off-line are conducted without oculomotor saccades.

WPM thus equals the online Rubin-Legge RSVP rate, which is also a nonsaccadic reading mode.

Off-line the consolidations or reviews are conducted by three concentric, actively functioning lexor systems and are undelayed by oculomotor saccades, as in the RSVP mode. Consider the sequential cortical maps with interpolations.

Thesis IV: A Shifting of Cortical Maps

Engrams can thus obtain alignment with the "polar-axial view."

The dynamics of Reading Online versus the *off*-line Lexication Dynamics.

When fluently reading Western texts online, the fixations are shifted to the right by two short saccades and by one long return saccade to the left across a stationary text.

However, when witnessing the *off*-line texts, the letters and lines in the cortical engram texts are themselves shifted leftward behind the stationary cortical lexors. Meanwhile, the body of the engram ascends slowly.

These mechanics, as observed since 2000 in Condition III, are seen in "off-line" lexication.

The slow up-drifting and leftward rotation of a lexical engram (while in Condition III undergoing scrutiny *off*-line) thus enables the continued operation of recursive nonsaccadic lexication. According to N. Haydn (2005), "We analyze a skew integrable map defined on a cylinder that models a shear flow." This dynamic may relate to my subjective findings.

Lexication *off*-line functions in the absence of motor saccades.

The slow up-drifting and leftward rotation of the engram (as may be witnessed *in limbo* status) enables a streaming operation across the three stationary, overlapping, concentric lexors.

Sequential Maps and Their Interpolations

D.A. Pollen (2009): "Fundamental requirements for primary visual perception."

With the voluminous incoming sensory data, there is integration of afferent and efferent signals across numerous cerebral areas, enabling feature-binding of entrainment and other coordinated and autonomous motor responses.

Thesis V: The Archived Compacted Multiple Engrams, or "ACME"

Multiple lexical engrams are retained intact but are minified: Henon maps?

Literacy is attained by

- Integrating of visual information from successive fixations. *(Jonides J., 1982)*
- Neuronal networks for induced 40 Hz rhythms. *(Jefferys J., 1996)*
- Sleep, off-line memory reprocessing. *(Stickgold R., 2001)*
- Memory consolidation in unsupervised periods during sleep, and by high-density storage, verbatim, of orthographic memory. *(IXP, 2000)*
- Archived Compacted Multiple Engram, ACME, 2003.

Thesis VI: The Active Default Brain

Awake or asleep, online or *off*-line, the bioergonomics of learning, of memory, and of recall mechanisms depend largely upon autonomous engrams and upon the "lexor circuitry," which is active in vigilance and also during non-REM short-wave sleep phases.

According to N. Axmacher (2008), "Memory Consolidation has been suggested to occur predominantly during sleep. Very recent findings, however, suggest that important steps in memory consolidation occur also during the waking state."

Routine typographic or editing corrections of a text made deliberately online have been witnessed as being replicated in some engrams. These "overnight consolidations" appeared to be conducted at accelerated speeds and were recently reconfirmed in September 2010.

Pelli and Tillman, in 2007, examined three scalar contributions to reading rates, attributable to the discernments of the Letters as symbols, of the Words as semantic, and of the Phrases as syntactic. They stated that "each reading process always contributes the same number of words per minute...accomplished by a complex process that contains three functionally distinct and separately modifiable parts, adding that the finding that these contributions to the reading rates are additive...is startling." (See Lexors)

However, these three lexors, which function as captors and couriers of three texton magnitudes, appear to operate concentrically, simultaneously, and congruently: as witnessed in engrams during *off*-line lexications. See Thesis VII below.

Thesis VII: Three Scalar Lexors, evidently operating in Synchrony.

This Engram Reading Mode functions with synchronous, concentric, three-*scalar* lexor systems. Each gestating engram is updated (i.e., remapped with each new fixation, until the closure of that visual reading epoch).

Teichert, in 2007, regarded "the scalar-invariance of receptive-field properties in primary visual cortex...[stating that] the neural mechanisms underlying these abilities are still poorly understood."

LEXICATION

On first confronting a text, the full extent of the visible text, paged or on screen, is promptly captured as a gist, a gestalt. As scanning progresses, three concentric afferent agents operate simultaneously and, with congruent synchronized summations, function as a triumvirate. Thus, the sequential views of the circular lexor's one-degree domain, of the five-degree circular domain, and those of the linear lexor with its three-by-ten-degree horizontal purview, jointly attain construction of the gestating engram.

These delegated afferent functions, implicit in vigilant online reading, may be rendered *explicit* in the *off*-line processing. As was witnessed in Condition III above, the spontaneous engrams emerged briefly during awakening arousals (as load- and recency-dependent) and are at times focally legible, *verbatim et literatim.*

Thesis VIII: Concentric Functionings of the Scalar Lexors

These sensory lexors with neural circuits of the three scalar magnitudes operate simultaneously during normal fluent reading activities.

Condition III above: Two concentric circular lexors have domains of a one- or of a five-degree subtense, while that of the linear lexor's horizon is a three-by-ten-degree subtense.

Textual details are furnished by these three concentric scalar lexors with their specific angular subtenses matched to the clusters of letters, words, or phrases: L, W, P. (The granular textons of B. Julesz, 1980.)

G. Legge, in 2007, using a flash card (a short block of text on four single-spaced lines) demonstrates lexical discernments, obtained thus in the field aperture of the five-degree circular lexor sensor, creating the sequential, nonoverlapping engrams.

Thesis IX: Semantic Cerebral Concatenations

Richard Sermon (1904): "…an engram…is a stimulus impression, reactivated by the recurrence of the prior conditions."

J. Born (2009): "The access sleep-dependent mnemonic consolidation requires memories to be encoded under control of prefrontal hippocampal circuitry…with the same circuitry."

B. Ermentrout (2009): "Effective connectivity between cells may arise in the form of correlations between noisy input streams. However, the inputs to real neurons may often be more accurately described as barrages of synaptic noise."

IXP (2000): "Engrams are triggered by flicker *toggled* between 10 to 40 Hz. concatenations."

Lexical Engrams as Witnessed in Their Recursion

I. Perception of Primary Visual Engrams is attained in Stochastic Noise

II. Secondary Engrams are triggered by Imposed Flicker, 10–40 Hz

III. Tertiary Engrams are processed overnight and daily in "Verbatim Consolidations"

IV. Quaternary Formats: The Archived Compressed Multiple Engrams (ACME)

Thesis X: Reading, Literacy, and Comprehension

Literacy is enabled by cognitive access to the archived retention, *verbatim et literatim*, of documents and by the early delegation to automata of many primary mnemonic functions. Awake or asleep, the bioergonomics of learning, of memory, and of recall depend ultimately upon autonomous and recursive neural functions.

According to E. Alvarez-Lacalle (2006), "Hierarchical structures induce long-range dynamical correlations in written texts...with presumptive-access to surrogate random texts identical to the original text." These "surrogates" have indeed been witnessed as facsimiled, *verbatim* engrams, together with *off*-line lexor activity evident in some ACME-archived displays.

ACME. The Archived Compacted Multiple Engrams. When glimpsed in the brief interludes between sleep and wake ("*in limbo*"), these engrams then display views of intact prior lexical inputs, subsequently witnessed in miniature in displays resembling "archived spreadsheets."

However, these engrams may be rendered explicit in three state-dependent conditions, as here asserted, and as documented since 2000. Richard Semon, in 1904, had described "engram reactivation by the recurrence of those energetic conditions which had ruled at the generation of that engram," and this statement broadly addressed the current issues of mnemonic entrainment and recalls elicited by triggered oscillations. The lexical engrams are found to be subjectively explicit and to be archived verbatim *et* literatim, thus enabling literacy and scholarship. The Engrame Reading Mode, as here presented, supports and extends current theory on lexical processes, memory consolidations, and the recalls in dormancy.

The Explicit Recursive Engrams

Subjective visual access to the Implicit PRS of Tulving and Schacter, 1990, is obtainable as recursive engrams, witnessed as explicit in conducive state-dependent conditions.

Recursive semantic engrams covertly serve cognition and literacy and may be rendered explicitly visible in three state-dependent conditions. During the casual online viewing of dynamic scenes or while reading text, these incoming visual patterns are autonomously packaged as the discrete engrams that are maintained with chronology and topologies intact. Their semantic contents are either of textural or scenic motions, or of graphic-lexical facsimiles.

Prior to the unsought phenomenal events of October 15–16, 2000, I was not familiar with reading neuromechanics or related theories. Since 2000, these phenomena became the major focus in exploration of the entoptic engram systems, which evidently constitute the substrates of literacy and of navigation.

Engrams experienced in the intrinsic cognitive field are anchored by the cognitive cyclopean fovea, a lynch-pin evident in all mnemonic packets engrams, and one in effect also during dream-image experiences—"the foveal confluence in human visual cortex," according to Schira, Tyler, Breakspear, Spehar, in 2009.

Summary

When lexicating *off*-line, there are no oculomotor saccades and no pages to turn. The visual effect is experienced as if a spiraling text is imprinted upon a vertical cylinder that is in a continuous levo-rotation at approximately 1 cps and that also attains a slow vertical ascent of negative three-degree subtense per second. This dynamic keeps the lines of the text in alignment with the concentric, autonomous, stationary sensory agencies. (See Pelli, 2007, "Agents and L, W and S.")

Visual Engrams

Thesis Statement: "Reading is an Engram-Dependent Function"

Abstract

Recursive Visual Engrams are the explicit substrates of the implicit PRS dual system of Tulving and Schacter, 1990. Visually explicit engrams are not prominent in the presence of homotopic online in the literature, for these engrams are imperceptible both in darkness and in the images. However, the recursive visual engrams in the orthographic and the scenic motion systems may be explicitly witnessed off-line in adaptive state dependencies.

Recursive Lexical Engrams are generated autonomously during saccadic reading and may thereafter be witnessed off-line in three conducive state-dependent conditions:

I) with closed, lit eyes, in stochastic resonance, perceptible after a five-second delay.
II) promptly seen with imposed flicker toggled between 8–40 Hz and in bright light.
III) spontaneously emergent engrams, witnessed in brief sleep-to-wake arousals.

At times in Condition III, these engrams are focally legible, *verbatim et literatim*, and may then equal the online RSVP reading rates, for both these processes are nonsaccadic.

The Dynamic Visual Engrams, discussed in the previous topic, are acquired autonomously while viewing natural fluid flows such as flocks of birds in flight, stirred coffee, vehicular traffic, falling snowflakes, and also the streaming texts on tickertape.

Complex motions and vortices are retained in tempero-spatial parity with their input sources, evident only in Condition I above.

The Engram Reading Mode as here presented contributes to current theories on lexical processing, memory consolidation, and perceptible engram recursions, triggered or spontaneous. A successful prearticulate infants' reading program is evidently reliant upon visual engrams (see Robert Titzer and D.L. Share, 2008, on intuitive learning), upon multimodal semantic inputs, both voiced and gestural when accompanying a displayed trigram, and upon subsequent cognitive access to the autonomously cerebrally imprinted lexical engrams.

Engram-Dependent Functions

The subjective manifestations of recursive semantic engrams are here attributed solely to the activities in the neocortex; whereas, mutual reciprocations and the roles of the hippocampus in memory storage and retrieval are detectable by electroencephalography.

The Recursive Lexical Engrams here characterized were acquired while routinely reading Times New Roman twelve-point print layouts, viewed at 75 percent from a distance of 50 cm.

Seen while awakening "in status limbo," findings of lexical activity are consistent with the generally suspected *off*-line consolidations. Caveat: The chances for encountering spontaneous engrams in a brief awakening period (of about ten seconds) are largely contingent upon the hours spent reading texts the previous day.

Subjective findings since October 2000. Levo-rotations and upward drifting of lexical engrams are perceptible during the accelerated off-line lexications. The familiar RSVP protocol had demonstrated a nonsaccadic reading mode. This is the mode also witnessed explicitly "off-line in limbo," for the engram texts rotate and shift vertically behind the static, immobile cortical lexors.

Slow spin and up-drifting of the intact texts, as witnessed in Condition III above, enables a continuous lexication as the static lexors discern the streaming text and attain continuous cognitive reading with no saccadic regressions.

S.M. Stringer (2004) in "Dissipative Dynamical Systems," states "Continuous attractor neural networks operate with localized packets…[and] such a network could, by learning, self-organize to enable the packets in each space to be moved continuously in that space by ideothetic (motion) inputs."

Since neither retinal nor cortical images can be scanned foveally by oculomotion, it is the informational "MAPS" themselves which thus may shift, enabling the off-line lexicational activities that are observed.

Off-line: No pages to turn, no oculomotor saccades.

Off-line: A vertical scroll, an endless up-winding flow, as in the RSVP mode.

The Subjective Findings Since October 2002

Levo-Rotations and Upward Drifting of Lexical Engrams Perceptible during Off-line "Consolidatory" Lexications

Engram texts may be witnessed during the off-line lexications. If texts were imprinted on a vertical conceptual cylinder, which is in a levo-rotation, and which attains a vertical ascent of two-degree subtense per second, this spiral delivery track could feed the now-stationary focal lexors: the three concentric, autonomous sensor agents, the foveola of one degree, the parafoveal of five degrees, and the gist of ten degrees, respectively the Letters, the Words, and the Phrases.

Up-drifting and rotation of the cortical engram itself thus enables the continuous operations of the three overlapping functioning lexors in streaming texts. This finding is perhaps consistent with a statement by N. Haydn in 2005: "analyzing a skew integrable map defined on a cylinder that models a shear flow." The subjective manifestations of explicit lexical engrams of textons and lexors prove supportive of current theories on reading skills and on literacy. The overt subjective manifestations of recursive semantic engrams are here attributed solely to the activities in the neocortex, while the reciproacting roles of the hippocampus in memory storage and its retrieval are polymodal and in locations elsewhere detectable by electroencephalography.

Other technology—sensor-chip compensation: One of the most advanced camera technologies is a system that actually moves the camera's sensor chip in a manner that counteracts a handheld camera's movement. All sighted and ambulatory species accomplish this same feat by making oculomotor and vestibular adjustments for the retinal intakes.

Entopic Landmarks

Pearce, Ida, M.D. May 1968. "Entoptic Visualizations and Impletion of the Blind Spots." *Arch Ophthal.* Vol 79.

Abstract

"The blind spot of Mariotte can be visible subjectively for as long as fifteen seconds at low luminances and also in the dark. When the adaptation of any part of the visual system does not match the stimulus, either a dark disc scotoma appears on the lit field, or a bright disc scotoma appears on a dark field. Often there is a border contrast phenomenon thought to be indicative of lateral inductance in the cortical fourth-order neurons. The scotoma is most slowly filled up at low luminances and is not perceptible with bright fields unless these have been preceded by prolonged dark adaptation. The initial impletion depends upon the contralateral retina; final "matching" impletion may be made by the complex cells, fifth order or higher, on signals from the ipsilateral terina, and at the same time suppression of the contra lateral "mismatching" signals."

By 2000, entoptic visualization of visual engrams and activities became apparent to me. Subjective access to the explicit substrates of the implicit PRS of Tulving and Schacter, 1990, is obtainable in three conducive state-dependent conditions. Recursive Semantic Engrams covertly serve cognition and literacy and may be rendered explicitly visible in three state-dependent conditions. During the casual online viewing of dynamic scenes, or while reading text, these incoming visual patterns are autonomously packaged as the discrete engrams that are maintained with chronology and topologies intact. Their semantic contents are either of textural or scenic motions, or they are graphic-lexical facsimiles.

The Engram Reading Mode as here presented relates to current theories of online lexical processing and memory consolidation and identifies autonomous recall triggers.

The successful Robert Titzer reading program for prearticulate infants was reliant upon multimodal semantic inputs, both voiced and gestural, and upon innate cognitive access to the autonomously generated lexical visual engrams that bear the inculcate orthographic symbols.

Recursive Lexical Engrams
Autonomy, Fidelity, and Longevities

Categorical Distinctions of Entopic Images
Iconic Memories: Implicit or Explicit?
Engrams: The Tern "Engram" Attributed to Richard Semon in 1904.
See D.L. Schacter, 2001, *Richard Semon and the Story of Memory.*
Mnemonic Storage Assumptions

Interdisciplinary Terminologies: Your Chunks = My Engrams
Mnemonic Research Methods
Defenders of Exact Subjectivism
Autonomy and Default Systems Including Central Pattern Generators, or CPGs

Categorical Distinctions of Entopic Images

"Classical early after-images:" These images are neither covert, nor motile, nor achromatic.

Photic after-images and the Pressure Phosphenes: Fluctuating over time with regard to apparent contrast and hue and with changes in ambient light, these retinal images *are* visible in the dark.

Eidetic imagery: S. Miller and R. Peacock (1982) described "long-lasting, percept-like experiences critically dependent on level of illumination and easily disrupted by after-coming visual stimuli" (as are engrams).

Imagery: A disclaimer: The autonomous engrams here reported are those witnessed and documented since 2000 by the present author, one of the many individuals unable to perform deliberative or guided "imagery." This faculty is claimed by some 80 percent of populations, with cortical activities validated by fMRI. Apparently, color is present in these "imagery" displays: the engrams here described are always achromatic and cannot be evoked voluntarily.

Bartolomeo P., 2009. An assessment battery for visual mental imagery and visual perception.

The **Scenic Motion-Memory** system is deemed innate; the **Orthographic system** inculcate.

Both these phenomena may be witnessed as explicit engrams and are deemed the substrates of the implicit PRS of Tulving and Schacter, 1990, with the Scenic or the Verbal mnemonic entities.

Results seem inconsistent with hypotheses postulating a strict correspondence between perceptual and imagery abilities. Engrams are not the products of imagery.

Semantic engrams appear in two functionally distinct mnemonic systems.
Engrams are retained in two functionally distinct mnemonic systems; one is innate, the scenic motion-memory system, while the other is an inculcate system for the orthographic symbols. Both these independent phenomena may be witnessed as explicit engrams; their functional segregations substrates are termed implicit in the PRS of Tulving and Schacter, 1990. The autonomous memory-traces of a single event, or of one page of text, constitute the mnemonic substrates of recognition, of recapitulations, and of motor entrainments.

The recursive motion-memory engrams exhibit decays over 120 seconds. The longevities of a covert lexical engram may be demonstrably explicit from minutes to days.

Iconic Memories: Implicit or Explicit?

Investigators referring specifically to a phenomenal "Iconic Memory" have included the following:

Coltheart M., 1983; Di Lollo V. and Dixon P., 1988; Sperling G., 1993; Billock V.A., 1997;

Tatler B.W. and Melcher D., 2007; Thomas L.E. and Irwin D.E., 2006; Sligte I.G., Scholte H.S. and Lamme V.A., 2008.

Various visual tasks, geometric or semantic, may be imposed upon the selected participants. Conclusions are then reached by the monitoring of their behavioral responses or by the use of electronic devices.

G. Borst and S.M. Kosslyn (2008) had their participants "scan over patterns of dots in a mental image...with images based on a just-seen pattern." There is some uncertainty as to what scanning entails in mental imagery.

S. Pinker (1998) suggests that "mental-image scanning is a process distinct from eye movements or eye movement commands" as is here affirmed.

E.D. Reichle (2006) describes, "effects of higher level language processing are not observed on eye movements when such processing is occurring."

V.A. Lamme (2009) states, "Recent fMRI studies have shown the existence of a form of visual memory that lies intermediate of iconic memory and visual short-term memory (VSTM), in terms of both capacity (up to 15 items) and the duration of the memory trace (up to 4 s)." Visual engrams are not subject to such brief temporal limitations, nor to the capacitance restrictions.

S.M. Stringer (2004). "Dissipative Dynamical Systems. MAPS." "Continuous attractor neural networks...operate with... localized packets...Such a network could by learning, self organize to enable the packets in each space to be moved continuously in that space by idiothetic (motion) inputs." Since neither retinal nor cortical images can be scanned foveally by oculomotion, it is these informational MAPS that shift during the off-line lexications, as is observed.

"Engrams" Literature from 1904

The term *engram* was coined in 1904 by Richard Semon, who defined engrams as "stimulus impressions...which could be reactivated by the recurrence of the energetic conditions which ruled at the generation of the engram." *Dorland's Medical Dictionary*, 1936, defined an engram as "a lasting mark or trace." *Concise Oxford Dictionary*, 1950: "Engrave: to impress deeply, as upon memory."

"In Search of the Engram" was published by K.S. Lashley in 1950. In 1930, Lashley had earlier written, "The facts of both psychology and neurology show a degree of plasticity, of organization, and of adaptation and behavior which is far beyond any present possibility of explanation." J.L. McGaugh, in 1966, commented, "Although this conclusion is still valid, the current surge of interest in memory storage offers hope that this conclusion may soon need to be modified."

"Further studies of the positive visual after-image," were reported by Padgham in 1957.

Glen Fry, in 1969, had discerned the "positive after-image" as having a composite nature and that it "demonstrated 120 second longevities after cessation of the stimulus." (Identical longevities of primary engrams were recorded by IXP on October 15, 2000.)

In 1967, O. Buresová, J. Bures, and M. Rustová had published papers entitled: "Acquisition and Retrieval of Visual Engrams," "Consolidation of Visual Engrams," and even "Interocular and Interhemispheric Transfer of Visual Engrams...in Callosotomized Rats"!

M. Coltheart, 1980, in "The Persistences of Vision": "Phenomenological neural persistence and informational persistence...are issues with which cognitive psychology has yet to come to grips."

S. Miller and R. Peacock, in 1982, wrote, "Eidetic imagery appears to be a long-lasting, percept-like experience which varies considerably in clarity and definition; its duration is critically dependent on level of illumination and its contents are easily disrupted by after-coming visual stimuli." Their five criteria closely match the engram parameters.

G.M. Long, in 1985, reported, "visual persistence from brief letters and pictures."

Tulving and Schacter, 1990, identified implicit learning by a perceptual representational semantic system, the PRS, suggesting that it has evolved to perform only ecologically valid computations.

D.A. Leopold, in 2003, noted that "by periodic closing of the eyes, perception can become locked or 'frozen' in one configuration for several minutes at a time…and that stimulus-selectivity of responses in primate infero-temporal cortex is stable across days and weeks of visual experience."

R. Blake, 1997: "…for subsequent deployment in cognitive judgments motion memory…necessarily maintained with some degree of accuracy."

S. Magnussen S, 1999: "…coupling to a memory store…series of parallel special-purpose perceptual mechanisms with independent but limited processing resources…"

D. Bibitchkov, 2000: "Pattern storage and processing in attractor networks…"

D.A. Pollen, 2003: "…recursive processing…the combined flow and integrated outcome of afferent and recurrent activity across a series of cortical areas…"

Y. Dudai, 2004: "…persistence, retrievability, and modifiability."

D.L. Schacter, 2004: "…a nonconscious form of memory… itemizing a previous encounter…"

E.T. Rolls, 2005: "…anticipates visual patterns with recall in high-fidelity…prolonged stable persistence after brief priming"

O. Jensen, 2006: "…multiple working memory items maintained by temporal segmentation."

Basole A, 2006, "…patterns of activity evoked by complex stimuli…best understood in the context of a single map of spatio-temporal energy."

S.M. Stringer, 2006: "…continuous attractor neural networks which can maintain a localised packet of neuronal activity representing the current state of an agent in a continuous space without external sensory input."

B.A. Wandell, 2007: "...principles at coarse scale which organize cortical spatial arrangements...and coordinate the timing of events in short-term memory."

M. Tsodyks, 2003: "...spontaneously emerging cortical representations."

M. Tsodyks, 2006: "...approximated versions of the orientation maps, emerging when the...network is presented with an unstructured noisy input."

O.L. White, 2004: "...sequences retrievable from the instantaneous state of the network."

N. Haydn, 2005: "...dissipative Henon map?"

R. Kempter, 2006: "...in long sequences apparently one order of magnitude below the capacity (multiple, miniaturized..." (See ACME.)

All the above criteria, especially those of Miller and Peacock in 1982, seem consistent with the findings of the explicit semantic engrams and their elusive qualities, state dependencies, and "reactivations."

Mnemonic Storage and Data Retrievals
Diverse Assumptions

D.C. Hall, in 1974, in considering mnemonic recall mechanisms, suggested that "eye movements scan iconic imagery." Such an ocular pursuit was deemed unlikely and was contended by B. Saskit in 1976, since the image moves with the eye.

E.D. Reichle, 2006, noted that "effects of higher level language processing are not observed on eye movements when such processing is occurring."

M. Coltheart, 1983, wrote, "The concept of a pre-categorical sensory memory remains tenable, with a transferring from iconic memory to a 'post-categorical' memory."

V. Di Lollo and P. Dixon, 1988, discerned "two forms of persistence in visual information: visible persistence, and the visual analog representation."

K.R. Gegenfurtner and Sperling, 1993, "...suggest that subjects must transfer information from a rapidly-decaying trace (iconic memory) to more durable storage."

V.A. Billock, 1997, postulates that "the site for very short-term (iconic) visual memory...could be Crick's reverberatory loop."

I.G. Sligte, H. S. Scholte, and V.A. Lamme, 2008, contend that "subjects can still access information from iconic memory because they can see an after-image (sic) of the display."

D. Melcher, 2007, cautions, "We must be very careful in our choice of scene stimuli, and in our interpretation of findings from the laboratory."

J.J. Fahrenfort, H.S. Scholte, and V.A. Lamme, 2007, note, "Re-entrant visual loops...several subsequent stages show an alternating pattern of frontoparietal and occipital activity, all of which correlate highly with perception."

Y. Dudai, 2010, "Signatures of Memory: Brain Co-activations during Retrieval."

Recalls are presumed effected by concatenations.

"Recognition" for a specific input by the reactivation of matching stored engrams predicates that the installation of each engram may be "*writ with a unique signature in broadband neuro-oscillations* and disseminated to a cohort of multiple-linked cerebral addresses."

J. Born, 2009, hypothesizes that access to sleep-dependent consolidation requires memories to be *encoded* under control of prefrontal-hippocampal circuitry "with the same circuitry controlling the subsequent consolidation during sleep."

Richard Semon, in 1904, had stated that "an engram could be reactivated by the recurrence of the energetic conditions which had ruled at the generation of the engram."

Concatenations

"Recognition" for a specific input by the reactivation of matching stored engrams predicates that the installation of each engram may be "*writ with a unique signature in broadband neuro-oscillations* and disseminated to a cohort of multiple-linked cerebral addresses."

The lexical *primary* **engrams** (in STM) are cognitively perceptible for up to 120 seconds and are shown in fMRI studies to reverberate in multimodal destinations, including the phonic and the motor speech centers.

Cognitive recalls are presumed aroused by harmonic concatenations from one or many circuits; recalls are rallied by "palpable hits" on the responding engrams.

Recognition of any online signal is attained by an alignment with various *prior c*ognitions and with sets of acquired biases (the previous "learnings.")

Interdisciplinary Terminologies
Neologisms, Terms, and Usages

Charles Dodson (Lewis Carroll) quotes the Red Queen telling Alice: "*When I use a word, it shall mean what I intend it shall mean.*"

The term *engram* was coined in 1904 by Richard Semon, who provided other neologisms. Engraphy refers to the encoding of information into memory; engram refers to the change in the nervous system—the memory-trace—that preserves the effects of an experience. Ekphory refers to the retrieval process (recursions) "*the influences which awaken the mnemic trace or engram out of its latent state into one of manifested activity.*"

D.L. Schacter, in 2001, drew attention again to Richard Semon and the story of memory: "Forgotten ideas, neglected pioneers…" a reissue of "The Stranger Behind the Engram."

S. Lehar, in 2003, comments on the "**Persistent Disparity** between the phenomenological or *experiential* account of visual perception…and the neurophysiological level of description of the visual system in computer models." P. Sinha, 2002: "Current artificial systems do not match the robustness and versatility of their biological counterparts."

The Visual Engram's "Sharp-bordered, Inviolate Domains."

Your Chunks = My Engrams!

Computer chunks are characterized here by L. Ceze:

As judged comparable to **visual engram** parameters, running comments follow:

A chunk in fine grain is an isolated atomicity, visible to other processes. Visual engrams have fixed boundaries, as if for read-only.

A chunk, during its installation, is impervious from outside its domain. Lexical engrams appear inviolable, but the dynamic motion engrams are subject to vestibular update interaction. (See Topic C7 "The Dynamic Engrams (Motion-Memory).") Heteromodal modification can thus be effected by an online vestibular input to an *off*-line motion engram. K. Patan, 2007: "Sequential order is maintained as a second chunk is initiated going to multiple posted addresses." Cached, recursive engrams likewise affirmed. (See ACME.) In-flight chunks can interleave. D. Zhao and L. Yang, in 2009, formulate the "Incremental isometric embedding…of high-dimensional data…using connected neighborhood graphs." In-flight chunks? These are the recursive subjective engrams. No evident interleaving is witnessed with lexical engrams. (See the a-tomicity argument and the claimed scalar parity integrity.) Yet some "compacted aggregations" are witnessed. (See ACME.)

Chunk order may be arbitrarily assigned for progressively higher performance through multiprocessor architecture, as for the engrams, being recycled for later literary consideration in recursions, which are explicit in status limbo.

Bulk invalidations are envisaged as a discarding of transaction memories.

Decay of the engrams is deterministic—or use it or lose it?

Discarding of transactional memories: In subjective findings of primary transactional motion_memories, the Dynamic Engrams decay by amplitude attrition, flat-lining over 120 seconds, whereas the lexical engrams are retained in chronologically ordered blocks over prolonged but indeterminate periods.

Visual Noise, the Bins, and the Aggregates

The following are clips culled chiefly from *Neural Computation*: H. Korn and P. Faure, 2003: Considered "the role of neuronal attractors in information processing, perception and memory... general issue of chaos as a possible neuronal code, and of the emerging concept of stochastic resonance." In the engram umbral views, perceptual thresholds of the primary engrams are lowered by the stochastic resonance. E. Alvarez-Lacalle and E. Moses, in 2009, note, "Slow and fast pulses in cultures of excitatory neurons demonstrate that proper levels of additive white noisy currents generate such pulses spontaneously."

Engram Acquisitions: Tempero-spatial limits are *encountered as empiric.*

T. Burwick, 2007: "...self-organized segmentation of two overlapping patterns...non-overlapping patterns can be simultaneously active with mutually different phases."

The superposition problem:

Merging: Successful online establishment in the context of one lexical engram, which is assembled with successive gists of one hundred overlapping concordant patterns made while reading the one page. (See geometric engram concordances and perceptual rivalries.)

C. Johansson and A. Lansner, 2007, note that "in order to achieve high storage capacity at least 20 to 30 pyramidal neurons should be compacted into a minicolumn and at least 50 to 60 minicolumns should be grouped into a hypercolumn." High storage capacity is seen with multiple minified intact lexical engrams. (See "ACME").

"What Are Textons?

Julesz, 1981: "The term 'textons' now refers to the fundamental micro-structures in natural images, and in texts, speech and music, and are considered as the atoms of pre-attentive human visual perception."

Song-Chun, in 2008, commented, "Unfortunately the word 'texton' remains a vague concept in the literature...for lack of a good mathematical model."

Mnemonic Research Methods

Methods Available in Mnemonic Research

Electro-technology: Behavioral Motor Responses: Subjective Evidence

According to N.C. Rust and J.A. Movshon (2005), "Traditional methods for exploring visual computations that use *artificial* stimuli with carefully selected properties have been, and continue to be the most effective tools for visual neuroscience. The proper use of Natural Stimuli is to test models based on responses to the Synthetic Stimuli, not to replace them."

However, the Engram Reading Process, which became evident during exposure to "Natural Stimuli" was not an experimental "testing of a model," nor was it undertaken upon any theoretical concepts or with prior familiarity with relevant literature.

Psychophysical Laboratory investigations of the human mnemonic systems commonly require that chins are immobilized and gazes fixed upon relatively impoverished or artificial stimuli "... despite the fact that primates actively examine the visual world by rapidly shifting gaze over the elements in a scene" (Rajkai C., 2008). In the study of reading skills, J. Pynte (2006) notes, "A methodology with five lines of text yields multiple inspection parameters." C. Pernet, in 2007, points out "the possible limitations of foveal stimulus presentation for drawing conclusions about natural reading...No neural effects appeared for LV Field primes, in line with the RVF preference imposed by the Western writing system."

Success in the learning of a set task may then be adjudicated by observing specifically linked responses, electronic or behavioral. Memories, when voluntarily recalled, are seldom claimed to reappear as topographic explicit images or to be achromatic, as are the autonomous recursive engrams.

Electro-technology, circa 2009. "The hemodynamic response tends to be more widespread in space and lasts longer in time as compared with the neuronal activity. Recursive signals have been identified in *single* neurons of invertebrates, rodents and primates, while the fMRI, VEP and other tools identify recursive activities at multiple human cerebral sites..." Roland and Gulyas, 1997, note that "the recognition of the same learned 'gross patterns' increased the regional Cortical BF in the 18 identically-located fields which overlapped activated during the learning, associated also with hippocampal activity."

S.F. Cappa (2006) states that "Some may then ask 'who cares about the localizations per se?'" B.A. Wandell (2007) asks who "cares about hodology, the white matter pathways in reading?"

Behavioral Motor Responses: Pavlov's concept of the dynamic stereotype was "imprints of reality." In its formulation in the 1930s, Pavlov's theory postulated the synthesis of conditioned reflexes, not associative chains of conditioned reflexes. Such Pavlovian reflexes indicate retentions of specific mnemonic data, yet "the recursive oscillations and synchronicities which enable cognition are found largely impervious to behavioral analysis" (Jack A., 2006). (See "Synesthesia and Synkinesia: Autonomous Vertical Ocular Saccades," Topic A3, concerning entrainment.)

Subjective Evidence of Autonomous Recursive Engrams

R.G. Jahn and B.J. Dunne, in 2007, in "Science of the Subjective," stated, "Over the greater portion of its long scholarly history, 'science' has relied at least as much on subjective experience and inspiration as it has on objective experiments and theories."

N.B. Turk-Browne, 2005: "For planned research on visual memory…it requires attention to select the relevant population of stimuli, but the resulting "learning" then occurs without intent or awareness."

L. Jiménez, in 2007, stated, "Evidence for chunk-learning can indeed be obtained before any systematic training, and thus surely reflect a pre-existent tendency rather than a learned outcome." Engrams are installed autonomously: "un-trained outcome events."

My own investigation of the Engram Reading Process began in 2000, prior to any familiarity with relevant literature or to theoretical concepts on mnemonic retention and in ignorance of the term *engram*. Bartels and Zeki, 2005, stated, "Natural viewing conditions can lead accidentally to validations of unsuspected correlations of phenomena, and may give impetus to new directions."

Subjective Phenomena
Defenders of "Exact Subjectivism"

O.J. Grusser, 1984, said, "Purkinje was a careful observer who proclaimed and exemplified the value of exact subjectivism," an approach we also may strive to emulate.

S. Lehar, 2003, commended "the primacy of subjective conscious experience rather than the modeling of the objective neurophysiological state of the visual system...which supposedly subserves that experience."

S. Dehaene, 2007: "Understanding the extent and the limits of non-conscious processing is an important step on the road to a thorough understanding of the cognitive and cerebral correlates of conscious perception."

R.G. Jahn and B.J. Dunne, 2007: "Any disciplined re-admission of subjective elements into rigorous scientific methodology will hinge on the precision with which they can be defined, measured, and represented...and on the resilience of established scientific techniques to their inclusion."

Empiric contributions of this **Engram Reading Mode** to theories of mnemonic networks derive from the unsought, serendipitous initiations and data-driven considerations. According to Bartels and Zeki (2005), "Natural viewing conditions can lead accidentally to validations of unsuspected correlations of phenomena, and may give impetus to new directions."

On searching recent literature on visual memory and on the neuromechanics of reading, none address as explicity the recursive engrams of orthographic or scenic motion dynamic systems. These elusive engrams remain imperceptible in the dark and in the presence of homotopic online images. These covert engrams are, however, continuously being installed autonomously, being reiterated autonomously, and proving explicit in some state dependencies.

Multimodal Engrams constitute the substrates of memory and of cognition in the short and in the long term, specific to each modality as audio-visual, verbal, motor, tactile, and olfactory, and any may be coincidentally linked. According to J. Vartianinen, T. Parviainen, and R. Salmelin (2009), "Regardless of the input modality, the Magneto Encephalography (MEG) studies indicate involvement of the middle superior temporal cortex in semantic processing from approximately 300 ms onwards."

The **Visual Engrams** are the most readily accessible for experiential study, encountered off-line in three adaptive state dependencies.

A. Bartels and S. Zeki, in 2005, stated, "An opportunistic approach...provides a powerful environment...to reveal connectivity in vivo."

The Lexical Engrams here reported are those acquired by the daily routine readings at a 45 cm distance with a standard font restricted to Times New Roman twelve points viewed at 75 percent. Displayed on a thirty-degree field of the monitor, each line of text may be captured in fluent reading by making three fixations per second. The two forward saccades average seven degrees, the return saccades approximately fourteen degrees.

Autonomy and the Default Systems

Paramount Significance of "Default Systems" and their Lynch-Pin Engrams

Two functionally independent mnemonoic systems act as the "captors, custodians and couriers" of the innumerable visual engrams, which in parallel dynamic and orthographic circuits jointly enable our reading skills.

The significance of the Engram Reading Mode relates to several issues: reshaping the binding problem of form and motion vision. According to M.R. Ibbotson (2007), "The apparent functional separation between these two paths has created a theoretical challenge. How can the information in the two systems be brought together across cortical space?"

5 **PRINT : STYLE . SCALE and LEGIBILITY : .**

. **Equivalent Acuity is posited in the awake # and in the lp * conditions**

With pt 12 Times in lower case , three letters measure 5 mm on the screen Five mm on the screen when viewed at 40 cms. subtends one degree, equaling the 3 micron foveola The fovea itself, some 1.5 mm in diameter, subtends ~ 5 degrees = on the screen, which may encompass 20 + letters . The parafovea extends a further 2 m around the foveal perimeter , the perifovea a further 1.5 mm . Reportedly the maximu reading rates are achieved with characters subtending 0.3 degree to 2 degrees. **Legge 198**

The constraints on the resolution of high spatial frequencies in normal vision # are also to be imposed in mental imagery * which therefore restricts how well fine details object ~~could be " imagined "~~ (**~~Finke 1985~~** ~~) ; however~~ " it may be ~~more difficult~~ to re high-resolution information in imagery than in perception" **Kosslyn , 1999** Although it is not ~~mentally- fabricated~~ imagery , ~~its facsimile~~ -copies engage ~~many networks~~ emp also in imagery , but the well-established parameters for on-line reading # may be com directly with some of the limited data obtained during the lp * state-dependent manife

The figure roughly illustrates the magnitudes of the three scalar lexors, which are concentric, not seriatim as here displayed.

While using standard texts with Times New Roman at twelve-point font size, single-spaced in the normal screen, and using print view, one line extending for 20 cm is read with three or four saccades. As may easily be verified when monocularly fixating on the center of the screen, the blind spots of Mariotte thus fall on the margins of a text as customarily viewed at approximately 45 cm, an arm's length.

Illustration of an engram witnessed September 12, 2001, at 6:00 a.m. in limbo. A spontaneous engram emergent in a brief sleep-to-wake arousal.

This 2001 portrayal (with an added, edited insert) depicted an engram of a two-column text. Overall illegible, yet within a one-degree circular lexor, there were perceptible three or four legible letters, strings streaming rapidly from right to left <<< and flickering out-of-phase with each other (at about 12 Hz) within the domain of a one-degree manifold of a 30-micron retinal foveola.

Three Sacades Each of 10 degrees places the Blindspots upon the foveal fixations of each succeeding saccade.

These now reported engrams appear in spatial parity, manifesting the visual data, which had been acquired during the reading epoch.

In the 30-cm normal view on screen, three lowercase letters and one space per cm subtends one degree when viewed at 45 cm, constituting one texton. Legge, in 1985, measured the reading rates for text scanned across the face of a TV monitor; maximum reading rates were achieved with characters of 0.3 degrees up to two degrees subtense. This subtense of one degree replicates that of the 300-micron retinal foveola, not that of the five-degree retinal fovea.

The contemporary note had described "a large area of coarse text with rare twin cursors…drifting downwards." By later interpretations the relative motions are recognized as an upshifting of the engram text.

Relative motions of the engram components may be witnessed during the off-line lexical processing. These motions, as subsequently witnessed in limbo, have indicated upshifting of the engram texts at about two degrees per second. The slow updrifting of the engram, and with a leftward rotating of the text, enables the continuous operation of the three stationary concentric, simultaneously functioning, cortical scalar lexors.

According to N. Haydn (2005), "analyze a skew integrable map defined on a cylinder that models a shear flow."

March 19, 2009. Attentive spotlights? The lexor phenomena are thus witnessed occasionally, explicit only in Condition III in a brief sleep-to-wake arousal.

The mechanism of the lexor, resembling "a bright soliton" increases the contrast of textons at locations within this lexor's one-degree boundary, matching the scope of the 30-micron retinal foveola.

The five-degree circular lexor, rarely seen, is scalar with the retinal fovea.

The lexor, with a three-by-ten-degree subtense (the interval between the optic nerve heads) is perceived as a stream darkening along a single line of an otherwise "illegible engram." Active only in the midthird of a line and only in the midthird a thirty-degree engram display field, this dimension equals the fixational domain in the online three-saccade fluent-reading performance.

Putatively, these functions are all registered simultaneously during each online fixation, and the compounded off-line data remain cognitively accessible, though generally imperceptible.

This engram reading paradigm supports and extends current findings and theories on lexical processing and on off-line memory consolidations.

The Attentive and the Default Functions
Of Educable Cellular Automata

DICTA

According to Turk-Browne (2005), "It requires attention to select the relevant population of stimuli, but the resulting learning then occurs without intent or awareness."

According to F. Pulvermuller (2006), "It is increasingly understood that lexical, semantic, and syntactic information can be processed by the central nervous system outside the focus of attention in a largely automatic manner."

According to Pelli (2006), "The full activation or ignition of specifically distributed binding-circuits explains the near-simultaneity of early neurophysiologic indexes of lexical, syntactic and semantic processing."

According to Jensen (2006), a cortical multi-item working memory buffer operates "in which theta and gamma oscillations have an important role…in recall."

According to D.L. Schacter (2009), "The resting brain is associated with significant intrinsic activity; fluctuations measured by functional magnetic resonance imaging. Their nature and function are poorly understood."

According to M.A. Goodale (2008), default networks act unsupervised during the SWS non-REM sleep periods, which support the task-related efforts in vigilance, and thus may act peremptorily in emergencies.

According to D.L. Schacter (2011), "Problem-solving simulations may allow default network regions to be coactive with task-positive network regions, without suppressing the contribution of either network."

Reading is an Engram-Dependent Skill
Contingent upon Autonomy and Default Systems

The term and the properties of the *engram* are attributed to Richard Semon, 1904; see D.L. Schacter's 2001 book *Forgotten Ideas, Neglected Pioneers: Richard Semon and the Story of Memory.*

Recursive Engram Findings since 2000. The recursive and covert lexical engrams, which are acquired while reading, are invisible in the dark, yet they may be witnessed as explicit in these three conducive conditions:

I) with closed, lit eyes in the umbral view, a stochastic-resonance effect.

II) with imposed flicker 4–40 Hz. O. Jensen, in 2006, portrays a cortical multi-item working memory buffer "in which theta and gamma oscillations have an important role…in recall."

III) as spontaneous images seen while awakening in limbo status. These findings are consistent with the generally suspected mnemonic consolidations, verbatim et literatim.

According to D.L. Schacter (2009), "The resting brain is associated with significant intrinsic activity; fluctuations measured by functional magnetic resonance imaging."

Semantic Engrams. Orthographic (Lexical) and the Scenic Motion (Dynamic) engrams are imperceptible in the dark or in the presence of homotopic online images, yet these engrams may be rendered explicit in the state-dependent conditions, as mentioned above.

These Recursive Lexical Engrams I first encountered in 2000, and the Dynamic Engrams were encountered later, in September 2001. Both these phenomena had emerged unsought and were then labeled "Illusory Images." Thereafter, I learned that both the term *engram* and the concept had been published earlier.

However, in the literature on visual memory and the neuromechanics of reading, few authors address as explicit these Recursive Orthographic and Dynamic Engram systems; putatively, these are the substrates of the Implicit PRS of Tulving and Schacter, 1990.

Engrams: "Specific Facsimile Memoranda"

Engrams constitute the substrates of memory and of cognitive functions in the short and the long term. They are specific to each modality: audio, visual, motion, verbal, corporeal motor, tactile, olfactory; any of these engrams may be temporally linked, most notably the audio-visual domains, both in the innate and the inculcate circuits.

Few readers find it remarkable that neither blinking nor saccades disrupts their view of the page before them, and many seem unaware of their own subarticular voicing as they silently read texts. With their eyes then closed in ambient light, few readers note that horizontal dark lines of printing may emerge to view following a four-second adaptive delay. These positive images are the primary lexical engrams.

Seen in a low salience and in coarse rendering in this shaded "umbral" state, stochastic resonance is obtained with the closed and illuminated eyelids. These recursive engrams may also be triggered by flicker from stroboscopic oscillations toggled between 10 to 40 Hz; they may also be elicited by simple finger flickering, thus by substituting a blank document screen in place of this currently read text, an engram may appear promptly when triggered by "Purkinje finger fanning."

With the two hands held open and close to the face and agitated past each other at a comfortable forty-five degrees and maximum speed, the oscillations may then approach effective range; the dark horizontal lines of an engram text can emerge promptly in fair contrast, the layout discernable. Purkinje, in 1823, had described "vierecke: large and small quadrilaterals...seen best by fixating a bright surface and waving the fingers rapidly before the eyes." Vierecke: The illegible words? An engram thus triggered may represent the layout of that text just read or facsimile of a text read earlier, for the longevities of these covert engrams exceed twenty-four hours.

Entoptic perceptions and functions of the Recursive Lexical Engrams are further detailed in Part Two of Memory Traces: Recursive Engrams: task-related mnemonic substrates that enable cognition, navigation, and literacy.

The online fluent reading with retention of semantic content is contingent upon cognitive access to the covert off-line Recursive Lexical Engrams. Pulvermuller, 2006, states, "Lexical information can be processed by the central nervous system outside the focus of attention...in a largely automatic manner." Subjective access to these covert recursive mnemonic images may be rendered explicit in three state-dependent conditions. It is concluded that "learning to read" is engram-dependent.

According to F. Pulvermüller, "The full activation or ignition of specifically distributed binding circuits explains the near simultaneity of early neurophysiologic indexes of lexical, syntactic and semantic processing." (See Pelli G., 2006, the L.W.S. categories.)

LEXICAL FUNCTIONING REQUIRES THE FOLLOWING:

- Incremental isometric embedding of textons while reading a page of text.
- High-dimensional data—with texts retained verbatim et literatim.
- Connected neighborhoods for multiple engram associations.

Literacy requires cognitive access to multiple archived engrams. It may be contended that during life the brain is never "at rest," dreaming in some phases of sleep and in other phases "consolidating" the very specific input imprints—the engrams of the previous day's experiences. These autonomous lexical processes that function during the day continue overnight during the short-wave sleep phases in the non-REM sleep.

Experiences of Explicit Recursive Visual Engrams

Chance encounters in 2000 initiated this ongoing study of phenomenal lexical engrams. Routine daily reading of Times New Roman, twelve points, viewed at 75 percent from a distance of 45 cm has, since 2000, served as the regular task, which has generated the covert autonomous lexical engrams.

Findings: Evidence of recursive visual memory systems is obtained following the free viewing of motile textures or structures and also of dynamic events. Following the saccadic reading of texts, and despite the number of saccades made (probably three per line, sixty per page), and regardless of the duration of each dwell point (probably of approximately 80 msec), the resulting engrams in recursion appear as intact photocopies, the achromatic facsimiles of the texts.

Protocols commonly require that chins are immobilized and gazes fixed upon relatively impoverished or artificial stimuli "despite the fact that primates actively examine the visual world by rapidly shifting gaze over the elements in a scene" (Rajkic C., 2008).

According to J. Pynte (2006), "In the study of reading skills a methodology with five lines of text yields multiple inspection parameters."

According to Bartels and Zeki (2005), "Natural viewing conditions can lead...accidentally...to validations of unsuspected correlations of phenomena, and may give impetus to new directions."

2000 July 22	Recursive Lexical Engrams
2000 Oct. 16	Autonomous Lexors
2001 Feb. 13	Engrams Triggered by Flicker
2001 Sept. 8	Spatial Motion-Memory, Dynamic Engrams
2003–2008	Lexical Archives with Shifting Engram Spreadsheets

Introspective Evidence Defenders of Exact Subjectivism

O.J. Grusser, in 1984, noted that "Purkinje was a careful observer who proclaimed and exemplified the value of exact subjectivism." This approach we may strive to emulate.

S. Lehar, in 2003, commends "the primacy of subjective conscious experience rather than the modeling of the objective neurophysiological state of the visual system...which supposedly subserves that experience."

S. Dehaene (2007): "Understanding the extent and the limits of non-conscious processing is an important step on the road to a thorough understanding of the cognitive and cerebral correlates of conscious perception."

R. Stickgold, in 2010, notes "hippocampus-dependent learning is accessible during non-REM sleep, and that hippocampus-mediated memory reactivation may be expressed, not only through neural activity in the sleeping brain...but also within concomitant subjective-experience."

E. Wamsley (2010): "Slow wave sleep was the only sleep parameter to correlate positively with declarative memory improvement. Introspective reports can provide a valuable window on cognitive processing in the sleeping brain."

The casual opportunistic viewings of dynamic events, of streaming fluids, and of turbulences generated the recursive motion-memory engrams, the second- and third-order motions here reported. (See Topic C7 "The Dynamic Engrams

(Motion-Memory)" in Part Two.) The normal saccadic reading of texts generated the lexical engrams.

In vivo Estimates: A familiarity with normal entopic structures, with retinal vasculature and with the Mariotte blind spot, has served for the subsequent evaluations and scaling of the landmarks of other entoptic patterns, notably the spiral and traveling waves, Topics D12 and D11, and for the Semantic Engrams, Topics B5 and C7.

Scalar parity, equated as the visual angle subtenses, matches the input sources. T. Teichert, in 2007, favors the hypothesis that processing in V1 supports scale-invariant aspects of visual performance. Intact visual engrams appear in spatial parity in terms of angular subtenses, thus putatively resonating throughout the sensorium.

An interesting exception is found in the case of the archived lexical engrams. ACME, the Archived Compacted Multiple Engrams, each minified by an order of magnitude.

Viewpoints Rigorous Subjective Experiences as Science?

Duke Elder, Systems of Ophthalmology XV Volumes: "Introspective experiments are usually considered second best in scientific investigations...but when carried out with rigor these have been of greater import in the study of human vision than the mass of objective undertakings." C.V. Mosby (1968): "Though predating the advent of fMRI technologies and in-vivo single-neuron recordings, his view remains pertinent." S. Lehar (2003) commends "the primacy of subjective conscious experience...rather than the modeling of the objective neurophysiological state of the visual system which supposedly subserves that experience."

"Experimentally unbiased and independent of hypothesis-driven, specialized stimuli...an opportunistic approach...relies... on natural brain dynamics."

Bartels and Zeki (2005): "Natural viewing conditions lead to particularly specific interregional correlations and thus provide a powerful environment to reveal anatomical connectivity in vivo."

R.G. Jahn and B.J. Dunne (2007): "Subjectivity has been progressively excluded from the practice of science, leaving an essentially secular analytical paradigm."

"Exact subjectivism by a very careful observer" was commended by Purkinje. This practice we also may strive to emulate.

Spatial, Dynamic, and Orthographic Mnemonic Engram Systems

These recursive semantic engrams reach explicit perception only in state-dependent conditions. Of the two functionally distinct mnemonic systems thus identified, one system briefly conserves "navigational data," which enables ambulation, while the other system maintains orthographic inputs with verbatim retention of the texts, which serves literacy.

E.T. Rolls (2005): "The verification of prolonged stable persistence of a visual pattern after brief priming and with recall in high-fidelity extends the horizons for theorists and experimentalists.

R.G. Jahn and B.J. Dunne (2007): Subjective data, while retaining the logical rigor, empirical and theoretical dialogue, and cultural purpose of its rigidly objective predecessor, have the following requirements:

- an acknowledgment of a proactive role for human consciousness (the biased prepared mind?)
- a more explicit and profound use of interdisciplinary metaphors (and defined specific terminology)
- more generous interpretations of measurability, replicability, and resonance
- a reduction of ontological aspirations.

Strict Automaticity?

Autonomous Neural Processes with deterministic responses may be detected in homo sapiens, in most animated species, and apparently also in the invitro cell cultures. L.M. Bettencourt (2007): "Functional structure of cortical neuronal networks grown in vitro. Despite their ex-vivo development, the connectivity maps derived from cultured neural assemblies are similar to other biological networks and display nontrivial structure in clustering coefficient, network diameter, and assorted mixing."

A.D. Milner and M.A. Goodale (2008): "Many action tasks have strict temporal constraints, which can only be met if the visual information is relayed directly to the motor system without first passing through a conscious decision-making process."

E.A. Di Paolo and H. Lizuka (2008): "Autonomous systems are the result of self-sustaining processes, constitution of an identity under precarious circumstances."

D. Harter (2005): "Being fast enough for use in real-time autonomous agent applications, use of a multilayer, highly recurrent model of the neural architecture of perceptual brain areas... develops action selection mechanisms in an autonomous agent."

T. Wantanabe (2000): "The architectural coupling between prey interception and retention has probably played a key role in both the macroevolution of orb web shape and the expression of plasticity in the spinning behaviors of spiders."

The Neuromechanics in Fluent Reading

The following metrics are obtained with a standard Times New Roman twelve-point font viewed at 75 percent. When read in print layout at 40 cm in normal horizontal linear reading(with three fixations per line), the two forward saccades each span ten to eight degrees and one return saccade of twenty degrees.

"The Writable Lexical Slate" in the brain measures approximately thirty degrees subtense: the 1-cm interval between the optic nerve heads, as is verifiable entoptically while witnessing the engrams. (See "Entoptic Visualization and Impletion of the Blind Spot." I. Pearce, *Arch. Ophth.*,1968.)

G.E. Legge and C.A. Bigelow (2011): "Does print size matter for reading? A review of findings from vision science and typography."

Findings in 2000–2012

With Times New Roman twelve points at 75 percent, in print layout, reading distances affect the optimal performances at 25 cm; reading takes four fixations at 50 cm takes three fixations, at 100 cm and needs only two fixations.

Evelyn Wood Speed-Reading: Instructions and Hand-Waving

"The insistence on hearing each word as you silently read restricts you to a reading speed based on how fast you can **hear** words...not how fast you can ***see*** words. Since you can only hear one word at a time...you can only read one word at a time."

These techniques were tested and proven at the University of Utah in 1959.

Investigators then suggested that phonemes and prosody experiences *limit* one's reading rates and are dispensable. This last statement may be fallacious; however, the successes claimed by devotees of these continued programs may be attributable to extended access to the cortical lexical engram, with the accelerated off-line lexication rate, which equals the speed of the RSVP. Ojanpaa, in 2002, noted that "in vertical lists, 4 to 5 words could be identified during a single fixation" (as may be obtained by the lexor with the five-degree subtense.) Reading columnar text rquires vertical not horizontal motions, thus speeding the process.

Thus, during each of the consecutive fixations made when reading online, those letters lying on the vertically adjacent lines above and below fixation are also legible to the five-degree foveal lexor and are eligible to be inserted autonomously into a currently gestating engram.

While reading a line of text the *vertically* adjacent graphic elements (textons) are acquired simultaneously with those in the *momentarily attended-to line* (*the signature line*). However, during normal fluent reading, the vector of attention is not distracted, nor is comprehension hindered.

All the images captured are retained in their dedicated engram gist in their preordained respective niches.

All three of the "concentric afferent lexical agents" evidently operate simultaneously and with congruent summations (see the L.W.S. of Pelli and Tillman, 2007) and as in the Engram Reading Mode: texton-lexors.

Text: Times New Roman, twelve points, single spaced, at 75 percent, 275 words on twenty lines, wpm at 50 cm, at 25 cm, or at 100 cm.

C. Schor (2008): "Implications for journal typographic design. Appropriate design modifications should allow greater user comfort and better performance."

Legge (2007): Oral reading 303 wpm for page text. Silent reading, up to 1,650 wpm? The average speed for text of an intermediate letter size was 1,171 wpm for RSVP. **As witnessed in limbo, the two *circular* lexors were of either a one-degree or of a five-degree subtense**; the linear lexor apparently had a three-by-ten-degree subtense.

Thesis: All three "*concentric* afferent lexical agents" operate simultaneously and with congruent summations. (See the L.W.S. of Pelli and Tillman, 2007.)

WPM: 2008 Reading Modes

Silent Prosody 450
Oral Prosody 260
Jabber, Fastest Aloud 350
Silent, Fastest 1,000
"Zipping" by Head Motion Only 1,200 with web layout
Fast-Finger Flow (Evelyn Wood claimed 1,000 wpm.)

Letter recognitions are made at fixations in the one-degree angular subtense; the words (standard Times New Roman twelve points) are recognized in the five-degree angular subtense and the phrases in three-by-ten-degree horizontal parcels.

The one-degree lexor captures graphemes less than three letters, as refined text. The lexor of five degrees captures words greater than ten letters (morphemes). The linear lexor spans three by ten degrees, capturing phrases greater than five words greater than 20 letters.

In normal horizontal linear reading, recognitions are made simultaneously by lexors in the one-, five-, and ten-degree subtenses. Thus, in fluent reading, the three concentric "afferent lexical agents" operate simultaneously and with congruent cortical summations. See the "surprise findings" by Pelli and Tillman, 2007; L.W.S. is consistent with the IXP Engram Reading Paradigm.

The autonomous engram systems are deemed the innate mnemonic assets, functional in most animate species—systems which presently support our own inculcate reading skills.

The standard pedagogic methods of reading instruction may be influenced by the Titzer enterprise "Your Baby Can Read" (1999) evidence, which prompts evaluations of current dogmas on learning, memory, and reading skills.

Visual Perceptions of Vestibular Signals (Topic F17)

"A Phantom Grid"

Experience: Projecting visual perceptions back into the environment. In darkness, luminous dots may be seen in a radial array; these are the manifestations of the Mandala template (Purkinje's Rosette). This array is a nonscannable visual experience, and on shifting the gaze this entoptic image moves through space as if projected from the retinas (cyclopean centering).

Head motions, either active or passive, may make it evident that a second more diffuse array or veil of fine punctate luminosities is following precisely the direction, speed, and extent of the head motions, a Phantom grid.

This creates the illusion of parallatic displacement between these two arrays, the more salient the polar lattice, the other a Cartesian grid.

Mechanism: Asynchrony between the slower retinal and the faster labyrinthine input contributes to an illusory perception: The Phantom Grid.

De Waele, in 2001, in "Vestibular projections in the human cortex" demonstrated that as early as 6 msec there is activation (at each of five identified multimodal cortical destinations) as shown with evoked potentials via tri-synaptic pathways from vestibular hair cells to these cortical destinations.

The eighth nerve's early-signal mechanism offers a possible advanced contextual placement for the online visual images, suggestive of one mechanism whereby ocular fixations are correctly assembled into engrams by the placement of letters into words, one fixational brick at a time, in cooperation each 300 msec with the oculomotor sensorimotor nerves.

It is supposed that this heteromodal binding, observed here in the dark, also operates constantly, though imperceptibly, in lit environments.

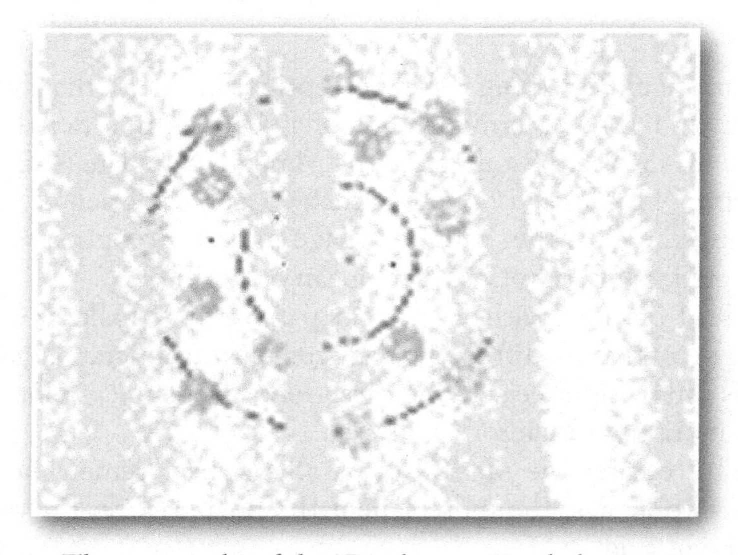

Schematic: The topography of the "Diaphanous Vestibular Bars" traversing the Polar Retinal Geometry.

Ambiguity and Biases (Topic F18)

Ambiguity: No images are instantaneously recognizable. All images are subject to algorithmic manipulations.

Rorschach ink blot tests. As interpreted by a viewer, the avowed semantic perceptions are assumed to be significant in that Gestalten are being generated by the underlying biases of the propositus and interpreted by an astute clinician. During the scanning of any real or pictorial object, as graphic or environmental, a definitive visual identification hinges upon the observer's current mind-set, expectations, prior experiences, concerns, and physical orientation. When in states of stress, inebriation, or paranoia or ecstasy, an explicit hallucination may be experienced and a false interpretation briefly entertained.

According to R.J. Snowden (2002), "In unfamiliar dim or dynamic surroundings, attention is drawn first to motion, or to the location of highest luminous contrasts; thereafter color clues, when available, may resolve ambiguity. Color signals are said to reach the human sensorium ahead of signals of motion or form." For aquatic primeval creatures, the early perception of spectral color or polarization may have had survival value.

According to C. Koch (2008), "Task-demands can immediately reverse the effects of sensory-driven saliency in complex visual stimuli."

Spatial Orientation: The initial perception of a horizontal conformity of the observer with the observed item is determinative in an interpretation of an ambiguous image. A tipping or tilting of the observer's head resets the algorithmic process, delaying a resolution.

State Dependency: In a contemplative reflective mood, the viewing of organic patterns can be enjoyed as artistically entertaining and invites illusory embellishments. This form of "false semantic morphing" is a permissive rather than a deliberative directed activity. The visual illusions that emerge over several seconds morph toward classic renderings of landscapes, human figures, and faces, rather than being degraded perceptually to simplified cartoon figures. For neonates, three-dots geometry is compelling and is autonomously sought after (also by adults) when seeking a human component to an enigmatic abstract display.

Heteromodal Inputs: Listening to music while watching a candle flame incites an apparent synchrony of flicker with music and exemplifies illusory correlations with causatively unrelated heteromodal signals.

Synesthesia and Synkinesia: Autonomous Vertical Ocular Saccades (Topic A3)

Auditory Patterns in Working Memory

"Language and music can be studied in parallel to address questions of neural specificity in cognitive processing. Prosody and music may overlap in the processes used to maintain auditory patterns in working memory." A.D. Patel (1998)

"Music increases frontal EEG coherence during verbal learning." D.A. Peterson (2007)

"Seemingly simple forms of associative learning are governed by multiple 'engrams' and by temporally dynamic interactions among these engrams and other circuit elements. Synapses, circuits, and the ontogeny of learning." P.S. Hunt (2007)

Engrams: Semantic patterns explicitly visible in state-dependent conditions; these data structures stored in attractor networks are accessible for entrainments and covert, recursive autonomous processing.

Fusion of visual and auditory stimuli during saccades: a Bayesian explanation for perisaccadic distortions. According to M.C. Morrone (2007), "During fixation, vision dominates and spatially 'captures' the auditory stimulus."

Listening to percussive music with both eyes closed and with a digital downward traction applied to immobilize both upper lids, rhythmic vertical ocular saccades may occur, as I first noted in 1985. When the eyes are open, the originally protective reflex up-gaze is largely inhibited in the interests of stable vision. The rhythmic vertical entrainment to percussion and to pitch as experienced with closed lids in downward traction represents an atavistic response to a possible threat to the eyes.

Music engenders ocular choreography, and audio-afferents can also modify entoptic visual images.

Zeki, in 1996, described "a direct demonstration (a V5 connection) of functional specialization within motion-related visual and auditory cortex of the human brain."

L. Glass (2010): "What is the origin of physiological rhythms? How do the rhythms interact with each other and the external environment?"

Audio-oculomotor Synkinesis: A reflex elevation of the globes synchronizes with percussive sounds and also to elevations of pitch. These effects are clearly visible to any onlooker and are palpable to the subject whose fingers are maintaining the downward lid traction. This reflex response can be demonstrated simply by rhythmic hand-clapping, or by prosodic speech; the most favorable subjects to demonstrate this phenomenon are dancers and musicians.

Percussive music: the "onset" is the most driving stimulus; a series of staccato stimuli progressively elevate the globe within the mechanical limits of the sensorimotor system, reaching a tetanic contraction at fast rates. Responses may fatigue within a minute yet after only three bars can be entrained by a rhythmic score and may then react in anticipation of the anticipated beat, even if this is withheld.

In a brief silence, an elevated ocular posture may be sustained briefly pending the next positive stimulus; otherwise, the globe will slowly descend to its neutral stance.

This myotonic activity is not volitional, but it is a permissive response. EOG tracings have shown that oculomotor upstrokes are active, ballistic, and myogenic, while the downstrokes usually represent an elastic restitutional return to a neutral horizon. This efferent ocular reflex response to sound is derived from the archaic startle response, initially the protective combination of blink and up-gaze.

An attitudinal gaze, the changing pitch of a sound, tends to an elevation or depression not only of the gaze but also of bodily posture. A change in audio volume may promote either advance or retreat along the personal Z axis, with innate extension or flexion of the limbs.

Natural hand gestures, body language, and choreography appear to demonstrate the innate more than any learned responses to patterns of sound.

Triggers for the Atavistic Vertical Saccades

Normally, a purposeful up-gaze is accompanied by bilateral elevation of the lids, synchronously programmed. But with downward traction made on both upper lids, thwarting lid elevation, a hyper-responsive status in the integrated vertical motion generator fires upon receipt of an uninhibited audio signal. Traction of only one lid (both eyes being closed) proves ineffective for this reflexive ocular audio-synkinesis. Entrained to selected music dynamics, the atavistic vertical saccades replicate a sequence or a series, stepped staccato impacts or smooth glissades, dependent upon the pacing of the audio stimuli.

Comment: The attribution of a bottom-up reflex ocular elevation to a startle-response origin is plausible, generalized when these vertical saccades are incited endogenously by silent private speech. However, atavistic, preserved roles for excitatory responses may provide an important mechanism in the formation of synapses and activity in neuronal networks.

Autonomous vertical ocular saccades are entrained by music and also by speech.

State Dependency. With both eyes closed and with a digital downward traction applied to both upper lids, patterns of acoustic sounds can induce vertical saccades. With patterned sound in the absence of visual input, oculomotor entrainments demonstrate the synchronous and modulated vertical saccades, an epiphenomenon, at times attaining 5 Hz.

Audio-motor synchrony and entrainments can be evaluated by finger-tapping and by EEG studies. According to U. Will and E. Berg, in 2007, "The tonic 2 Hz. maximum corresponds to the optimal tempo identified in listening, tapping synchronization, and event-interval discrimination experiments."

Entrainment to periodic acoustic stimuli demonstrates time coupling between rhythmic sensory input and motor output. Natural hand gestures, body language, and music performance itself demonstrate both the biological innate resonances and the acculturated responses to patterns of sound as conveyed in varied regional dance styles; entrainment is also heard in group applause.

R. Jackendoff, in 2005, noted the "specific way that tonal pitch contours can evoke patterns of posture and gesture." It may be observed that a changing pitch tends to an elevation or a depression, not only of bodily posture, but it also occurs as the vector of the autonomous vertical ocular saccades as here described.

Empirical evidence of oculomotor entrainment to patterned sounds is simple but is state dependent, both for its installation and for the tactile experiences of these recursive motions, which are palpable for the participant and visible to an observer. This autonomous motor activity matches the acoustic dynamics of speech and of musical input.

According to E. Brattico (2006), "Musical scale properties are automatically processed in the human auditory cortex. The relational properties of the musical scale are quickly and automatically extracted by the auditory cortex...even before the intervention of focused attention."

Entrainment to periodic acoustic stimuli demonstrates autonomous time coupling between rhythmic sensory input and vertical ocular motion.

With patterned sound, in the absence of visual input, in status, the oculomotor entrainments demonstrate an epi-phenomenon; the synchronous and modulated saccades are vertical only and exceed the 2 Hz optimal fingering tempo, at times attaining 5 Hz. (Paganini, it is said, could finger at 12 Hz.)

T. Iwasaki and M. Zheng (2006):"Sensory feedback mechanism underlies entrainment of a central pattern generator to mechanical resonance. Robust entrainment results from coupling of a lightly damped mechanical system with the reciprocal inhibition oscillator. The six pairs of ocular muscles represent the most versatile of all human complex motor coupling systems, enabling precise conjugate binocular oculo-versions over 100 degrees to a tolerance of a few seconds of arc, controlled by afferent feedback."

Furthermore, without the developmental acquisition of sequential and dedicated automata, including place-saver devices, neither the unified perception of a scanned field nor the acquisition of literacy would be attained.

In response to abrupt or shifting locations of environmental noises, reflex oculomotion acts autonomously to direct the visual attention appropriately. Patterned audio input, if possible being of semantic significance, attracts high-level attention and exploratory oculomotion to serve vision.

If a novel audio input is not on mental file as a known pattern, its distinguishing from random noise turns upon a third iteration, which then marks an event of potential significance.

According to M.A. Goodale (1998): "The dorsal 'action' stream projecting from primary visual cortex to the posterior parietal cortex provides flexible control of more ancient subcortical visuomotor modules for the production of motor acts, generally on-file protective."

These motor acts may include this quasi-physiological epiphenomenon, autonomous vertical saccades. Simple linear readouts from generic neural microcircuits (engrams) that send feedback of their activity to the circuit can be trained (entrained, polymodal audio/visual/tactile) using identical learning mechanisms to perform quite separate tasks of decision making and generation of subsequent motor commands.

Ancient Subcortical Visuomotor Modules

The eyeball is protected by several innate reflexes, which include a startle reflex with its simple lid closure. A forced closure of the lid may be accompanied by an ocular up-gaze: Bell's phenomenon, 1823.

The following autonomous ocular motions are protective:

Spontaneous blinking rates maintain the precorneal tear film. According to Adler (1992), "In a stable environment, these blinks can average every 3 seconds, with durations of 300 msec, a stereotyped dynamic."

M. Iwakai (2005): "Spontaneous blinks produce small eye movements directed down and inward, whereas slow or forced blinks were associated with delayed upward eye rotations (i.e., a Bell's phenomenon."

The Reflex Eye Blink. Percussive sounds commonly elicit a blink response, an audio-motor reflex. Simple blink responses, widely studied with Pavlovian conditioning, are reported with varied sensory modalities in many species and in contexts including psychopathology and pharmaceutical interventions.

Classical Blink Conditioning (Cason, 1922).

M. Bangert (2006): "The eye-blink reflexes trained to patterned sound are responsive to absent tones in repetitive stimuli."

This indicates an entrainment involving branches of the eighth and seventh cranial nerves, and of other somatic motor systems. Entrainments subsist upon memory-traces, formulated as engrams.

According to J.H. Freeman (2007): "The medial auditory thalamic nuclei and their projections to the pontine nuclei are components of the auditory conditioned stimulus pathway in eye blink conditioning."

Bell's Phenomenon, 1823. This reflex elevation of the globes is induced with the eyes closed in primary position and with pressure applied to prevent levator lid action. M. Takagi, in 1992, noted, "Prolonged eyelid closure...accompanied by tonic eye elevation replicating an ancient excitatory response."

M. Iwasaki, in 2005, noted of EEG potentials: "Spontaneous blinks produced small eye movements directed down and inward, whereas slow or forced blinks were associated with delayed upward eye rotations" (i.e., Bell's phenomenon).

In the formation of synapses and activity in neuronal networks, evolutionary-preserved roles for ancient excitatory responses may provide important mechanisms. In the following perverse autonomous ocular motions here detailed, these actions are not directly related to semantic vision processing, but to pattern per se. According to T.Q. Gentner (2006): "The capacity to classify sequences from recursive, centre-embedded grammars is not uniquely human. Recursive syntactic pattern learning by songbirds."

Speech and music travel upon a common path. Music and speech patterns entrain vertical ocular saccades. Vertical ocular saccades can be entrained by imposed percussive sounds, by simple rhythmic hand-clapping, and by the patterns heard in music and in normal human speech.

P.S. Hunt (2007): "Simple forms of associative learning (entrainments) are governed by multiple 'engrams' and by temporally dynamic interactions among these engrams and other circuit elements."

State Dependency. An experience of these audio-motor entrainments is evident only with both eyes closed and downward traction of the upper lids. These motor acts are an entrainment to periodic acoustic stimuli. The time coupling between rhythmic audio input and vertical ocular saccades constitutes a physiological epi-phenomenon, generally inhibited.

The acoustic experiences that can incite vertical saccades include the following:

1) Hearing music with both eyes closed and with a digital downward traction applied to the upper lids.
2) Listening, in status, to the speech of others. The consonants and the prosody in the delivered speech patterns can trigger elevations of the globes, modulated in frequency and amplitude.
3) Speaking aloud: Saccadic synkinesis is again present when the words are self-articulated.
4) Whispering the same phrase (at 20 dB?).
5) Hummed speech mode, although lacking in expletives, is effective. See "Processing Prosodic Boundaries in Natural and Hummed Speech," by A.K. Ischebeck (2007), an fMRI study.
6) Reminiscing, silently rethinking, articulating a phrase or a brief comment, and then mentally reiterating the identical phrase in silence can also replicate the previous patterned excursions of the globes. While these "virtual audio stimuli" continue in silence, the oculomotor sequences may be videotaped or recorded by EOG. Responses for up to

thirty seconds may accompany some ten iterations when made using a three- or four-second silent phrase, as this small packet is recapitulate in "private speech."

7) Inventing a novel phrase, mentation, and thinking up but not expressing a novel phrase does not immediately trigger this oculomotor response, but after several iterations an oculomotor entrainment may continue for many cycles.
8) Silent, steady counting by seconds or packaged as phrased triplets greater than fifteen packets. With silent recitation of the alphabet, phonemes, up to twenty, give inconsistent results. That unvoiced, private speech can also generate this autonomous epi-phenomenon has seemed remarkable.

Private Speech

Reading and Prosody. When silently reading, most people seem unaware of their own nonarticulated inner voice as audio background, yet most recognize this when it is brought to their attention. When reading, the ocular muscles execute approximately eight-degree subtense saccades at approximately 4 Hz, arguably by voluntary, top-down, goal-oriented, target-directed, horizontal saccades.

According to D. Girbau (2007), "Neurocognitive research analyzes the neural activity of individuals during a variety of task settings, including spontaneous and instructed overt and inner private speech use, sub-vocal verbalizations, and overt and silent reading."

As witnessed in the lid traction epi-phenomenon, the saccades are not goal oriented nor target directed, and the motions are strictly vertical. These saccades respond autonomously to detailed dynamics within each specific phrase while hearing music or speech and also while speaking, and more surprising, during the silent rethinking of the earlier phrase, matched vertical saccades arise, apparently programmed by the sensitized CGS's audio-motor engram.

Zeki, in 1996, dealing with prosody, described "a direct demonstration (a V5 connection) of functional specialization within the motion-related and the auditory cortex of the human brain."

According to P. Cornelissen (2006), "With magneto-encephalography data obtained during silent reading, the face motor cortex and the cerebellum, typically associated with speech production, emerged as densely connected components of the network." This finding implicates a covert entrainment initiated in the Private Speech Mode (PSM). In PSM, the prosody translates into a "baton effect" witnessed with covert vertical saccades.

Entrainments, it is suggested, are driven by recursive engrams of limited longevities.

The explicit visual motion-memory engrams (SMME), as witnessed, have iteratedfor up to 120 seconds, supposedly as recursions between hippocampal buffer statuses and multiple effector and sensory cortical destinations.

Learning Synapses, C

"However, by 2005 newsprint format was requiring only vertical saccades needed by the narrow columns of ten degrees sub tense, thus omitting horizontal saccades and accelerating reading speeds."

ircuits, and the Ontogeny of Learning

P.S. Hunt (2007): "Seemingly simple forms of associative learning are governed by multiple 'engrams' and by temporally dynamic interactions among these engrams and other circuit elements synapses, circuits, and the ontogeny of learning."

D.A. Peterson and M.H. Thaut (2007): Music increases frontal EEG coherence during verbal learning.

Recursive Engrams

Visual sensory entrainments establish explicit off-line recursive engrams.

Lexical-Visual Engrams and Afferent Visual Motion Engrams

Explicit visibility of covert visual engrams is obtainable with closed illuminated eyes witnessed explicitly as idiosyncratic images with verbal specificity.

The visual motion engrams may be installed or encoded in low resolution at approximately 3–8 Hz. These autonomous, achromatic primary stereo engrams are packaged in three- to thirty-second blocks that reiterate and attenuate over a visible longevity of some 120 seconds. (See C7 "The Dynamic Engrams (Motion-Memory).")

According to W.M. Joiner (2007), "An internal clock for predictive saccades is established identically by auditory or visual information" and linked to recursive motion memory-traces/engrams, and "for subsequent deployment in cognitive judgments… a motion-memory system must necessarily be maintained…and with some degree of accuracy" (Blake R., 1997).

Both these explicit visual engrams with idiosyncratic and verbal specificity compare in performance with the implicit Perceptual Representation Systems (PRS) described by Tulving and Schacter in 1990 in "Priming and Human Memory Systems."

The autonomous saccadic audio entrainments posit another recursive motor-memory engram system, comparable but briefer than those of the autonomous visual motion engrams, wherein each system retains highly specific content, with engram manifestations in state dependencies.

Senkowski, in 2007, reported, "Temporal synchrony of the unisensory components of an audio-visual stimulus with audio-visual stimuli presented with onset asynchronies ranging from -125 to +125 ms." Acoustic experiences that entrain these (atavistic) vertical saccades may conform to similar temporal parameters, 4 Hz.

Speech Patterns. The prosody, stresses, and percussive consonants are reflected verbatim by vertical ocular saccades. These vertical oscillations are synchronized with an online phrase, whether this is delivered by others or is self-articulated, voiced or unvoiced as whispering, or even while silently rethinking this same now entrained phrase. Silently thinking a novel phrase can also generate this response, but only after a few mental repetitions of the novel phrase. According to P. Cornelissen (2006), "With magneto-encephalography data obtained during silent reading, the face motor cortex and the cerebellum, typically associated with speech production, emerged as densely connected components of the network."

A treatise on facial muscles and "habit responses" appears in "Expression of Emotion in Man and Animals" by Charles Darwin, 1806.

The Extrinsic Eye Muscles

For each globe there are six anatomically paired extraocular muscles. In various gazes a muscle may function cooperatively as agonist or antagonist. Intrinsic efferent and afferent innervation is via the III, IV, and VI cranial nerves, with modalities of extrinsic sensory input also via visual II and auditory VIII, vestibular proprioceptive, facial VII, and a V trigeminal motor connection.

According to Buisseret, in 1985, "Extrinsic eye muscles respond to both proprioceptive and visual impulses" (and also to audio and mentation signals.) "The afferent pathway for the proprioceptive sensory system initially accompanies the motor fibers of oculomotor III, IV, and VI nerves. Proprioceptive afferent signals also project on two cerebellar cortex regions, one in the VIth and VIIth lobules of the posterior lobe vermis."

W.M. Joiner (2007): "An internal clock for predictive saccades is established identically by auditory or visual information."

R. Blake (1997): "A motion memory system for subsequent deployment in cognitive judgments must necessarily be maintained...and with some degree of accuracy." The cursive motion memory-traces visual engrams approach these criteria with a fair degree of idiosyncratic specificity.

The Highly Idiosyncratic and Verbal Specificity of the Autonomous Audio Phenomenon, as of the Visual Engrams

P.A. Luce, in 1998, said that many theories of spoken word recognition assume that lexical items are stored in memory as abstract representations. However, "representations of spoken words in memory are veridical exemplars that encode specific information, such as characteristics of the talker's voice" (Goldinger, 1996).

It appears that the vertical oculomotor saccades entrained to speech in umbral status are experienced as precisely modulated excursions, finely tuned to the dominant percussive patterns in music and driven also by prosody and phrasing patterns of normal articulated speech, and as I detected in 2007, they are also responsive to the unvoiced mental iterations of those same phrases. Indeed, after several cycles, there may be response to a novel and previously unuttered phrase.

A.R. Bradlow (1999): "Effects of talker, rate and amplitude variation on recognition memory for spoken words showed an advantage in recognition memory for words repeated by the same talker and at same speaking rate (but no advantage occurred for the amplitude condition.)"

U. Will and E. Berg (2007) in "Entrainment to Periodic Acoustic Stimuli": "The synchronization responses in the delta range may form part of the neurophysiological processes underlying time-coupling between rhythmic sensory input and motor output."

T. Isa and Y. Kobayashi (2004): "Express saccades…visual input appears to be transformed into motor output via a 'short-loop,' brainstem-mediated pathway."

Q. Yang (2006): "Aging deteriorates the ability to trigger regular volitional saccades but does not deteriorate the ability to produce express type of saccades." This finding also suggests that the so-called express saccades may be innate and autonomous.

P. Miller and X.J. Wang, in 2007, demonstrate analytically and computationally the exponential dependence of stability on the number of neurons in a self-excitatory network: "an interesting ramping temporal dynamics as a result of sequentially switching an increasing number of discrete, bistable, units." Entrained to the music's dynamics, the vertical saccades may replicate, in series or sequence, stepped staccato impacts or smooth glissades.

The Longevity of Engrams

During the visible longevity of an explicit recursive visual-motion engram, Dynamic Engrams, the amplitude of the signals attenuates, but frequencies remain unchanged. These engrams retain coherent pictorial representations with complex confluent motions and with state-dependent visible longevities for a maximal 120 seconds. (See Topic C7 "The Dynamic Engrams (Motion-Memory)." Both visual motion and the primary graphic engrams attenuate by 120 seconds. (See also "The Lexical

Engrams" (within Topic B5), "Recursive Lexical Engrams (Mnemonic Recall)."

Audio Oculomotor Engrams. The lid traction, state-dependent saccadic response has been demonstrated only up to forty-five seconds during the iterations of brief phrases.

P. Miller and X.J. Wang (2007): "[S]tability...relates to the stochastic fluctuations in self-sustained neuronal autocatalytic systems."

There are, for each globe, six anatomically paired extraocular muscles. Each muscle may function as agonist and/or antagonist in various gazes. Intrinsic efferent and afferent innervation is via the III, IV, and VI cranial nerves, with modalities of extrinsic sensory input also via visual II and auditory VIII, vestibular proprioceptive, facial VII, and a V trigeminal motor connection mechanism in the formation of synapses and activity in neuronal networks.

The six extraocular muscles, variously yoked as agonists and opponents, are innervated by III, IV, and VI cranial nerves, with modalities of extrinsic sensory input via visual II and auditory VIII, vestibular proprioceptive, facial VII, and trigeminal connections.

Related Pathology: Echolalia and Autism

"However, by 2005 newsprint format was requiring only vertical saccades needed by the narrow columns of ten degrees sub tense, thus omitting horizontal saccades and accelerating reading speeds."Systemic delays in attenuating audio-motor engrams may promote the utterances in echolalia, and the perseverate, echoic actions of autism. Evolutionary-preserved roles for excitatory responses may provide important mechanisms in the formation of synapses and activity in neuronal networks.

Western Reading Patterns Employ Horizontal Saccades

However, newsprint format was requiring only vertical saccades needed by the narrow columns of ten degrees sub tense, thus omitting horizontal saccades and accelerating reading speeds. According to R. Engbert (2001), "Mathematical models of eye movements in reading suggest a possible role for autonomous saccades." In the normal print view, one line in Times New Roman twelve points may be processed (i.e., read) with three saccades, enabling some 300 msec for the processing of each phrase (perception itself requires less than 40 ms exposure).

Structured upon innate and developmental phonemes and upon prosody, skillful reading becomes a rhythmic quasi-autonomous, inculcated activity. Numerous protocols and parafoveal studies approach reading saccades as predictive, "planned," reactive, recursive, and express.

Prosody, in reading three saccades, two each of approximately ten degrees to the right and one twenty-five-degree return sweep to the left, enables three fixations. With familiar content, this pattern can become an entrained, frequency-locked performance. When reading silently yet with this (inevitable?) nonarticulated inner voice, the phrasing of approximately fifteen to twenty letters in a well-written text may often seem coincidental with the reader's visual spans. Until brought to their attention, many readers seem unaware of their own nonarticulated inner voice.

Reflex Horizontal Saccades Pterygoid Muscle Activities May Also Incite Oculomotor Synkinesis

D. Goodisson and L. Snape (2000): "The jaw-winking syndrome involves synkinesis of the pterygoid muscles and levator palpebrae superioris, whence the eyelid retracts with mandibular movements...a motor branch of the trigeminal nerve innervates the external pterygoid muscle."

M.K. Bhutada (2004): "The lateral pterygoid muscle participates in fine control of horizontal jaw movements."

Synergy Findings, with Closed or Half-Closed Eyes

Repeated horizontal saccades of a full 180 degrees, deliberately initiated, may entrain auxiliary lateral jaw motions in the same directions.

As implemented by alternating pterygoid activity, the laterally gliding motions at the tempero-mandibular joint can incite brisk horizontal saccades of the globes' horizontal saccades...and can incite horizontal jaw movements, lateral gliding motions at the tempero-mandibular.

Repeating midline protrusion of the jaw may be accompanied by ocular convergence.

The progenitors of pterygoid muscles and of the ocular horizontal muscles shared an ontological lineage.

According to M.K. Bhutada (2004), "...the lateral pterygoid muscle participates in fine control of horizontal movements."

The evolutionary-preserved role for excitatory responses provides an important mechanism in the formation of synapses and activity in neuronal networks.

"Extraocular muscle precursors form tightly aggregated masses that en masse cross the crest mesoderm interface to enter periocular territories...progenitors of individual muscles do not establish stable, permanent relations with their connective tissues until both populations reach the sites of their morphogenesis within branchial arches or orbital regions."

S.M. Barlow (2006): "Central pattern generation, subserving oro-facial motor behavior, can be modulated via descending and sensory inputs of infrastructure for suck, respiration, and speech."

O. Hoshino (2007): "When a sound stimulus is presented to a subject, the auditory cortex first responds with transient discharges across a relatively large population of neurons, showing widespread-onset responses..." (including motor responses).

Recursions in Excitable Media

Repeated horizontal saccades of a full 180 degrees, when deliberately initiated, may entrain lateral jaw motions in the same direction: a pterygoid activity.

State-dependent persistence: Audio and Motor Engrams may persist unduly in echolalia and autism?

K. Hadano (2007): "...echolalia in patients with cerebral infarction...unable to repeat sentences longer than those containing four or six words (the size of an auto-engram phrase)."

Audio and Motor Engram Persistence in Echolalia and Autism

H. Hara (2007): "So-called 'idiopathic' autism, which exhibited no major complications before diagnosis is well-known as one of the risk factors for epilepsy...> 25%."

L.A. Femia (2002): "...co-morbid conditions such as epileptiform discharge during sleep and sleep disorders...it is possible that autism could involve a breakdown in consolidation processes."

R. Legenstein and W. Maass (2008): "On the classification capability of sign-constrained perceptrons. The perceptron also referred to as McCulloch-Pitts neuron, or linear threshold gate."

Glossary

Acculturations: The inculcate, learned, synaptic adaptations that autonomously conserve and process innumerable archived lexical engrams, enabling literacy.

ACME: Archived Compacted Multiple Engrams; in high storage.

Acquisition: The point-for-point tempero-spatial installation of a visual primary engram.

After-Images: Temporary photic scotomas; bistable resets; such are not the semantic engrams.

Anchor: Cyclopean fixator, a singularity, "the Grand Vizier," the cortical fixator.

Angular Subtense: That of a primary engram matches its online originator's subtense.

Archives: Numerous inviolate engrams reappear regimented in simultaneous displays.

Associative Learning: Is contingent upon recursive engrams.

Auto-catalytic Processes: Bioenergetic mechanisms that sustain mnemonic recursions. (Bandwidths for the lexical recursions can toggle across 4–40 Hz.)

Automata: "...neuro-biologically inspired agents" *(Cortes J.M., 2007).*

AVBP: Autonomous Visual Brain Phenomena.

Caesura: A "cloture" for a currently nascent engram.

Centering: The emplacements of textons during the generation of primary engrams.

Chunk: For the motion engram effective with epochs of three to thirty seconds; for the text, one page equals a chunk.

Coarse Rendering: The standard visual rendering in the primary engrams. *(See Malone, B.J., 2007; Bar M., 2006.)*

Cognitive Slate: Thirty-degree platform; equals three ten-degree saccades, which serve for acquisition of most lexical engrams.

Coherent Lexical Processing (CLP): One lexical engram: a clip! or a chunk!

Consciousness: The small candle on the big cake.

Consolidations: The putative (often nocturnal) *verbatim* processing of lexical engrams.

Covert Images: Implicity data, but potentially explicit in some state-dependent manifestations.

Default: An autonomous response not solicited by an investigator.

Encoding: Sensory coding, decoding, and representations—unnecessary and troublesome constructs? Prefer transforms? *(See Halpern, B.P.., 2000.)*

Engrams: Autonomous memory-traces of a single event or of a one-paged text; sequenced images in time-locked memory packages.

Entoptics: Endogenous visual images.

Entrainment: See Topic A3 and music.

Event Horizon: The tempero-spatial constraint limiting the view to any one discrete engram.

Explicit: Visually perceptible, online or off-line.

Feature: A portion, or one aspect, of the whole item.

Fixator: The cortical focal singularity linked to, responding to, or fed by bifoveal inputs; cognitive ideocentric anchor.

Fovea: Equals five-degree subtense; the parafovea approximately ten-plus-degree subtense.

Foveola: Equals 300 micron on retina; its lexor subtending one degree, capturing three to four characters of Times New Roman twelve points.

Gists: The "bare bones" of structures or gestures, as seen in a coarse primary engram.

Grapheme: Equals three to four textons at Times New Roman twelve-point text (equals one-degree foveola, which equals lexor i).

Hodology: The pattern of white matter connections between cortical areas tractography.

Holographic: Mnemonic Dynamic Engrams may be experienced as if in four dimensions.

Hub: Cyclopic center; axial pole; fixator; foveola; umbo; pivot point; belly button (singularity?).

Hypnopompic or Hypnagogic States: Between sleep and wake; *in limbo*.

Idiosyncratic Images: Those fabricated by imagery and in dreams (distinct from this default autonomous engram phenomenon.)

Implicit: "…with unconscious inference."

Inculcate: Sensory or motion patternings imposed into a mnemonic neural system, manifested as engrams (i.e., learned).

Inlaying: The emergence, witnessed entoptically in limbo, of an unrelated five-degree engram arising centrally upon a thirty-degree lexical engram. For reading, the angular subtense of an engram matches its online subtense, which equals the perceptual lexical field.

Installation: The autonomous generation of an engram, synonymous perhaps with priming in some protocols, or with adaptation, as with MAE.

Inviolate: Intact borders; nonpenetrable.

Learning: Here ascribed simply to autonomous imprintings in engram systems.

Lexication: Reading, notably specifying as performed off-line.

Lexigrams: Arbitrary graphic stimuli; symbols, ciphers, code.

Lexor i: Of approximately one-degree diameter; this clocks three to four tokens at 7–10 Hz.

Lexor ii: Of approximately five-degree diameter; encompasses words less than ten letters.

Lexors: Autonomous sensory agents, explicit with the rereadable engram texts *in limbo*.

Limbo: A tag for the hypnopompic state; the transitional mental status experienced in the borderland between sleep and wake.

Linear Lexor iii: Narrow horizontal strip less than ten degrees that engages phrases (approximately five words/twenty letters) in the midthird of a line of print on a thirty-degree engram.

MAPS: The shiftable network domains that enable the functional readouts between compilations of several archived engrams.

Memorable: Any suprathreshold sensory stimulus.

Memorized: In an engram, a status identified by reiteration or recall.

Merging: See the successful online establishment in the context of one lexical engram assembled with one hundred overlapping concordant patterns.

Noise: See "Stochastic Resonance."

Off-line: Includes the intrinsic, subjective versions of verifiable engrams.

Online: Verifiable perceptions of extrinsic items.

Oscillons: See "Rotating Spiral Wave Systems."

Parafoveal Functions: Extend to the ten-degree isopter, there bracketed by the ON blind spots.

Parallel Processing: Lock-step reading of contemporary spatial motion with lexical engrams.

Parity: For reading that the angular subtense of an engram matches its online subtense, commonly the thirty-degree perceptual lexical field.

Perception: Any salient image; term as used elsewhere: a percept that equals online versus an image that equals off-line.

Phase-Space: The domain of a dynamic or lexical engram.

Plasticity: A term here eschewed.

Primary Engrams: Discrete packets with recursive visible longevity of approximately 120 seconds.

Priming: *(Tulving, 1990)* Here understood as the acquisition or installation of a primary engram; elsewhere, an imposed hint or clue initiating a search for a target or tip. Priming the "input" by a researcher; otherwise the normal ubiquitous engram intakes.

Proscenium: Concept here: The only perceptual platform that supports the onstage visual perceptions, awake or asleep, and is anchored backstage by the cortical fixator, a singularity.

Purple Entoptic: A phenomenal annulus evoked at 36+ Hz. (See Topics E13, E14, and E16.)

Qualia: The chroma, contrast, luminance, or resolution of images.

Reading: Goal-oriented discriminations with scalar hierarchies.

Reading Skills: Proceeding from letter recognitions up to vectors of attention.

Recall: A solicited, deliberate search for an associated archived memory.

Replay or Recursion: Here taken as an autonomous reiteration of an engram.

Resonances: Stochastic and specific imposed oscillations.

Scanning: Mental-image scanning is distinct from eye movements or commands.

Semantic: With informational content.

Stochastic Resonance: Noise has a role in the umbral viewing of primary engrams.

Task: A continuous viewing of one congruent event within one epoch of time.

Tertiary Engrams: Those seen *in limbo.*

Texton: In a texture, local conspicuous features are called textons (*Julesz, 1980).*

Texton Gradients: A symbol, cipher, number, or letter (class i); a word (class ii); a phrase (class iii). These are visible contiguous elements covering areas or linear segments. Elements such as tiles, ciphers, symbols, alpha-numeric letters, words or line segments; texture elements.

Translocation: Upshifting of an engram across the intrinsic perceptive field.

Transparency: Simultaneous perception of overlying images; no perceptual interfacing is seen between exo- and endo-homotopic images.

Triggers: Flicker at 10–40 Hz can evoke a visible replay of lexical engrams.

Umbral View: Perceptions obtained with closed illuminated lids in stochastic resonance.

Veridical Images: The online perceptions of confrontational visual fields.

Versions: The renderings of engrams as modified by ambient light, noise, flicker, and time.

Zipping: Rapid reading of long lines of a web layout, achieved by making only smooth head motions.

Bibliography of Suggested Reading

Topic A1 - Ocular Pressure Phosphenes and the Microcirculation Dynamics

Purkinje, J.E. Beobachtungen und Versuche zur Physiologie der Sinne. Beitrage zur Kenntnis des Sehens in subjectiver Himsicht. Bd. I Prague. (1823).

Helmholtz, H. von. "Physiological optics." quoted J. Opt. Soc Am. (1924).

Bernstein, P. Raman. "Detection of macular carotinoid pigments in human retina." *Investig.*

Ophth. and Vis.Res. (1998). 39 2003-20011.

Brown, C. and J.W. Gebhard.Visual Field Articulations in the Absence of Spatial Stimulus. pp 188-99.

Bill, A. and G. Sperber. "Control of retinal and choroidal blood flow." *Eye.* (1990). 4 [Pt2]: 319–25.

Kaczurowski, MJ. "Chromatophores of the human eye." A.J.O. (1963). 56; 766–85.

Eisner, A. "Multiple components in photopic dark adaptation." J Opt Soc Am A. (1986). May; 3[5]:655–66.

Wang, L., M. Kondo, and A. Bill. "Glucose metabolism in cat outer retina. Effects of light and hyperoxia." *Investig. Ophth & Vis. Science.* (1997). 38[1]; 48–55.

Brown, C. et al. "Entoptic perimetry screening for central diabetic scotomas and macular edema." *Ophthalmology.* (2000). 107; 4, 755–759.

Buerk, D.G., C.E. Riva, and S.D. Cranstoun SD. "Frequency and luminance-dependent blood flow and K ion exchanges during flicker stimuli in cat optic nerve head." *Invest. Ophthalmol.*

Topic A2 - Hypnagogic Images

Chen, W. "Human primary visual cortex and LGN activation during visual imagery." *Neuroreport.* (1998). Nov 16: 9(16): 3669–74.

Di Lollo, V, J.T. Enns, and R.A. Rensink. "Competition for consciousness among visual events: the psychophysics of reentrant visual processes." *J Exp Psychol Gen.* (2000). Dec: 129(4): 481–507.

Fendrich, R. and P.M. Corballis. "The temporal cross-capture of audition and vision." *Percept Psychophys.* (2001). May: 63(4):719–25.

Jouny, C., F. Chapotot, and H. Merica H. "EEG spectral activity during paradoxical sleep: further evidence for cognitive processing." *Neuroreport.* (2000). Nov 27:11 (17): 3667–71.

Kleinschmidt A., C. Buchel, C. Hutton, K.J. Friston, and R.S. Frackowiak." The neural structures expressing perceptual hysteresis in *visual letter recognition*. Perception can change nonlinearly with stimulus contrast, and perceptual threshold may depend on the direction of contrast change." *Neuron* 2002 May 16; 34(4):659-66 Tachistoscopic recall ??

Lewald J, W.H. Ehrenstein, and R. Guski. "Spatio-temporal constraints for auditory—visual integration." *Behav. Brain Res.* 2001 June; 121(1-2):69-79. Bimodal neurons such as have been demonstrated by neurophysiological recordings in midbrain and cortex.

Makeig S., T.P. Jung, and T.J. Sejnowski. "Awareness during drowsiness: dynamics and electrophysiological correlates." Can. J. Exp. Psychol. 2000 Dec. 54(4):266-73. Episodes of nonresponding lasting about eighteen seconds.

Maquet P. "The role of sleep in learning and memory." *Science* 2001 Nov 2; 294(5544):1048-52.

Poldrack, R.A. and M.G. Packard. "Competition among multiple memory systems: converging evidence from animal and human brain studies." *Neuropsychologia.* 2003; 41(3):245-51.

Roland, P.E. and B. Gulyas. "Visual memory, visual imagery, and visual recognition of large field patterns by the human brain: functional anatomy by positron emission tomography." *Cereb. Cortex.* 1995 Jan–Feb; 5(1):79-93 The extravisual networks mediating storage, retrieval, and recognition differ, indicating that the ways by which the brain accesses the storage sites are different.

Seigneux P., S. Laureys, X. Delbeuck, and P. Maquet. "Sleeping brain, learning brain. The role of sleep for memory systems." *Neuroreport* 2001. Dec. 21; 12(18): A111-24.

Shams, L., Y. Kamitani, and S. Shimojo. "Visual illusion induced by sound." *Brain Res. Cogn. Brain Res.* 2002. June: 14(1):147-52. The temporal window of these audio-visual interactions is approximately 100 ms.

Spiridon, M. and N. Kanwisher. "How distributed is visual category information in human occipito-temporal cortex? An fMRI study." *Neuron* 2002. Sept. 12; 35(6):1157-6565.

Steriade, M., I. Timofeev, and F. Grenier. "Natural waking and sleep states: a view from inside neocortical neurons." *J. Neurophysiol.* 2001 May. 85(5):1969-85. High firing rates in the *functionally disconnected state of slow-wave sleep,* we suggest that neocortical neurons are engaged in processing internally generated signals.

Vinson, M . "Control of spatial orientation and lifetimes of scroll-rings in excitable media." *Nature.* 1999. April 3: 386(6624): 477-80.

Walker, M.P., C. Liston, J.A. Hobson, and R. Stickgold. "Cognitive flexibility across the sleep-wake cycle: REM-sleep enhancement of anagram problem solving." *Brain Res. Cogn. Brain Res.* 2002 Nov. 14(3):317-24

Watanabe K. and S. Shimojo. "When sound affects vision: effects of auditory grouping on visual motion perception." *Psychol. Sci.* 2001 Mar. 12(2):109-16. A genuine cross-modal effect.

Topic A3 - Music and Ocular Choreography

Calvert, G.A . "Evidence from fMRI of crossmodal binding in the human hetero-modal cortex." 1: Curr. Biol. 2000. Jun 1; 10 (11):649-57.

Chen, W. "Human primary visual cortex and LGN activation during visual imagery." *Neuroreport.* 1998. Nov.16:9(16):3669-74.

Di Lollo, V., J.T. Enns, and R.A. Rensink. "Competition for consciousness among visual events: the psychophysics of reentrant visual processes." *J. Exp. Psychol. Gen.* 2000 Dec. 129(4):481-507.

Farroni, T. "Configural processes at birth; evidence for perceptual organization." *Perception* 2000: 29(3):355-72.

Fendrich R., P.M. Corballis. "The temporal cross—capture of audition and vision." *Percept Psychophys.* 2001. May: 63(4):719-25.

Galati G., G. Committeri G, J.N. Sanes, and L Pizzamiglio. "Spatial coding of visual and somatic sensory information in body-centred coordinates." *Eur. J. Neurosci.* 2001 Aug; 14(4):737-46.

Ferreira, C.T. "Separate visual pathways for perception of actions and objects: evidence from a case of apperceptive agnosia." *J. Neurol. Neurosurg. Osychiatry.* 1998 sSeot; 65(3):382-5.

Grusser, 1995 ref; Gerhardt, M. "A cellular automaton model of excitable media including curvature and dispersion." *Science.* Mar 30 1990 1563-6.

Harrer, G. and H. Harrer. "Music, Emotion and Autonomic Function in "MUSIC and the BRAIN "1997 Eds . Critchley and Henson 1997.

Hubbard, T.L. "Synesthesia-like mappings of lightness, pitch and melodic interval." *Am. J. Psychol.* Summer; 109(2):219-38.

Jacobs, K. "Audio-visual synesthesia: sound-induced photisms." 1: *Arch. Neurol.* 1981 April; 38(4):211-6.

Knill, D.C. "Contour into texture: information content of surface contours and texture flow." *J. Opt. Soc. AMA Opt. Image Sci. Vis.* 2001 Jan; 18[1]:12-35.

Lewis, T.J. "Self-organized synchronous oscillations in a network of excitable cells coupled by gap junctions." *Network.* 2000 Nov; 11(4):299-320.

Logthetis, N.K. "Functional imaging of the monkey brain." *Nat. Neurosci.* 1999 Jun; [6]:555-62.

Maess, B. "Musical syntax is processed in Broca's area: an MEG study." 1: *Nat. Neurosci.* 2001 May; 4(5):540-5.

Mendola, J.D. et al. "The representation of illusory and real contours in human cortical visual areas revealed by fMRI imaging." *J .Neuro. Science.* 1999 Oct 1; 19(19): 8560-72.

Miller, S. "Evidence for the uniqueness (sic) of eidetic imagery. *Percept Mot. Skills*. 1982 Dec; 55[ ]:1219-33.

Shams, L., Y. Kamitani, and S. Shimojo. "Visual illusion induced by sound." *Brain Res. Cogn. Brain Res.* 2002 Jun; 14(1):147-52. (100 msec offset)

Stickgold, R. "Replaying the game: Hypnagogic images in Normals and Amnesiacs." *Science*. 2000 Oct13; 290(5490):350-3.

Speilman, A.J. "Intracerebral hemodynamics probed by near in-fra-red spectroscopy on the transition between wakefulness and sleep." *Brain RES*. 2000 Jun 2; 866(1-2):313-25.

Tassi, P. and A. Muzet. "Defining the states of consciousness." *Neurosci. Behav. Rev.* 2001 Mar; 25(2):175-91.

Tenenbaum, J.B. "A global geometric framework for non-linear dimensionality reduction." *Science*. 2000 Dec 22; 290[5500]:2319-232.

Wilson, H.R. "Evolving concepts of spatial channels in vision; from independence to non-linear Interactions." *Perception*. 1997; 26(8):936-60.

Burkell, J.A. and Z.W. Pylyshyn. "Searching through subsets: a test of the hypothesis." *Spat. Vis.* 1997; 11(2):225-58.

Fendrich, R. and P.M. Corballis. "The temporal cross-cap-ture of audition and vision." *Percept Psychophys*. 2001 May; 63(4):719-25.

Galati, G, G. Committeri, J.N. Sanes, and L. Pizzamiglio. "Spatial coding of visual and somatic sensory information in body-centred coordinates." *Eur. J. Neurosci.* 2001 Aug; 14(4):737-46.

De Waele, C. "Vestibular projections in the human cortex." *Exp. Brain Res.* 2001 Dec; 141(4) 541.

Loose, R. and T. Probst. "Angular Velocity, not acceleration of self motion, mediates vestibular-visual interaction." *Perception.* 2001; 30(4)511-18.(4):719-25

Pylyshyn, Z.W. "Situating vision in the world." *Trends Cogn. Sci.* 2000 May; 4(5):197-207.

Tyler, C.W. and L.L. Kontsevich. "Stereoprocessing of cyclopean depth images: horizontally elongated summation fields." *Vision Res.* 2001. Aug; 41(17):2235-43.

Galati, G. "Committeri coding of visual and somatic sensory information in body-centred coordinates." *Eur. J. Neurosci.* 2001 Aug; 14(4):737-46.

Topic B5 - Lexical Engrams

Anagnostou, E. and W. Skrandies. "Effects of temporal gaps between successive fixation targets on discrimination performance and evoked brain activity." Neurosci. Res. 2001. Aug; 40(4):367-74.

Anstis, S. and A. Ho. "Nonlinear combination of luminance excursions during flicker, simultaneous contrast, afterimages and binocular fusion." *Vision Res.* 1998. Feb; 38(4):523-39.

Antonietti, A., B. Colombo, and M.J. Perez-Fabello. "Lack of association of computer use and ability with spontaneous mental visualization." *Percept Mot. Skills.* 2002. Feb; 94(1):182-4n.

Avidan, G. and M. Behrmann. "Correlations between the fMRI BOLD signal and visual perception." *Neuron.* 2002. May 16;.34(4):659-66. (hystersis)

Baars, B.J. "Attention versus consciousness in the visual brain." *J. Gen. Psychol.* 1999. July; 126(3) 224-33.

Barlow, H.B., R. Narasimhan, and A. Rosenfeld. "Visual pattern analysis in machines and animals." *Science.* 1972. Aug. 18; 177(49):567-75.

"Letter recognition in a 3x3 matrix." *Neural. Netw.* 2000. Oct-Nov; 13(8-9):941-52.

Bartolomeo, P. "The relationship between visual perception and visual mental imagery: a reappraisal of the neuropsychological evidence." *Cortex.* 2002 Jun; 38(3):357-78 "... *recent reports of patients showing double dissociations between perception and imagery abilities challenged the perception-imagery equivalence.*"

Brockmole, J.R. and R.F. Wang. "Switching between environmental representations in memory." *Cognition.* 2002. April; 83(3):295-316.

Brockmole, J.R., L.A. Carlson, and D.E. Irwin. "Inhibition of attended processing during saccadic eye movements." *Percept Psychophys.* 2002. Aug; 64(6):867-81.

Brockmole, J.R., R.F. Wang, and D.E. Irwin. "Temporal integration between visual images and visual percepts." *J. Exp. Psychol. Hum. Percept. Perform.* 2002. Apr; 28(2):315-34.

Buchel, C. "The predictive value of changes in effective connectivity for human learning." *Science.* 1999. Mar 5; 284(5407) 158-41. *Activation in specialized cortical areas decreases with time.*

Burr, D.C. "Electro-physiological correlates of positive and negative after-images." *Vision Research.* 1987; 27(2):201.

Brunel, N. "Slow stochastic Hebbian learning of classes of stimuli in a recurrent neural network." 1: *Network.* 1998 Feb.9(1):123-52.

Cave, K.R., S. Pinker, L. Giorgi, C.E. Thomas, L.M. Heller, J.M. Wolfe, and H. Lin. "The representation of location in visual images." *Cognit. Psychol.* 1994 Feb; 26(1):1-32.

Chen, W. "Human primary visual cortex and LGN activation during visual imagery." *Neuroreport.* 1998 Nov. 16:9(16):3669-74.

Chen, C.C. and C.W. Tyler. "Spatial pattern summation is phase-insensitive in the fovea but not in the periphery." *Spat. Vis.* 1999; 12(3):267-85.

Cornelissen, P. et al. "Coherent motion detection and letter position encoding." *Vision Res.* 1998 July; 38(14):2181-91.

Del Guidice in 1998: *"Networks that behave as palimpsest. "The illusory documents we have seen as persistent lexical texts evidently are sustained by such a network."*

Di Lollo. "Inverse- intensity effect in duration of visible persistence" *Psychol. Bull.* 1995 Sept.; 1 118(2); 223-37.

Esposito F. et al. "Spatial independent component analysis of functional MRI time-series: to what extent do results depend on the algorithm used?" *Hum. Brain Mapp.* 2002 July; 16(3):146-

Finke, R.A. and H.S. Kurtzman. "Mapping the visual field in mental imagery." *J. Exp. Psychol. Gen.* 1981 Dec.; 110(4):501-17.

Gerling, J. and L. Spillman. "Duration of visual afterimages on modulated backgrounds; post- receptoral processes." *Vision. Res.* 1987; 27(4): 521-7.

Gulyas, B. "Neural networks for internal reading and visual imagery of reading: a PET study." *Brain Res. Bull.* 2001 Feb; 54(3):319-.

Grossberg, S. "The link between brain learning, attention and consciousness." *Conscious Cogn.* 1999 Mar; 8(1):1-4.

Hobson, J.A. and E.F. Pace-Schott. "The cognitive neuroscience of sleep: neuronal systems, consciousness and learning." *Nat. Rev. Neurosci.* 2002 Sep; 3(9):679-93.

Irwin, D.E. "Lexical processing during saccadic eye movements." *Cognit. Psychol.* 1998 Jun; 36(1): 1-27.

Irwin, D.E. and J.R. Brockmole. "Mental rotation is suppressed during saccadic eye movements." *Psychon. Bull. Rev.* 2000 Dec; 7(4):654-61.

Jacobs, A. "Eye -Movement control in visual search: How direct is visual span control?"

Perception & Psycho. Physics. 1986 39[1]47-58.

Jefferys, J. "Neuronal networks for induced 40 Hz. rhythms." *Trends Neurosci.* 1996 May; 19 (5):202-8.

Jones, M.J., P. Sinha, T. Vetter, and T. Poggio. "Top-down learning of low-level vision tasks." *Curr. Biol.* 1997 Dec 1; 7(12):991-4.

Jonides, J. "Integrating visual information from successive fixations." *Science.* 1982 Jan; 215[4529]:192-4.

Kosslyn, S.M., K.E. Sukel, and B.M. Bly. "Squinting with the mind's eye: effects of stimulus resolution on imaginal and perceptual comparisons." *Mem. Cognit.* 1999 Mar; 27(2):276-87. "It is more difficult to represent high-resolution information in imagery than in perception."

Kennedy, A. "Parafoveal processing in word recognition." *J. Exp. Psychol.* A 2000 May; 53(2):429-55.

Kleinschmidt, A., C. Buchel, C. Hutton, K.J. Friston, and R.S. Frackowiak RS. "The neural structures expressing perceptual hysteresis in visual letter recognition." *Neuron.* 2002 May 16;34(4):659-66.

Kosslyn, S.M., K.E. Sukel, and B.M. Bly. "Squinting with the mind's eye: effects of stimulus resolution on imaginal and perceptual comparisons." *Mem. Cognit.* 1999 Mar; 27(2):276-87.

Laureys, S., P. Peigneux, F. Perrin, and P. Maquet. "Sleep and motor skill learning." *Neuron.* 2002 Jul 3; 35(1):5-7. Review Cyclotron Research Centre, University of Liege, Sart Tilman, 4000 Liege, Belgium.

Legge, G.E., S.J. Ahn, T.S. Klitz, and A. Luebker. "Psychophysics of reading—XVI. The visual span in normal and low vision." *Vision Res.* 1997 Jul; 37(14):1999-2010.

Chung, S.T., J.S. Mansfield, and G.E. Legge. "Psychophysics of reading. XVIII. The effect of print size on reading speed in normal peripheral vision." *Vision Res.* 1998 Oct; 38(19):2949-62.

Legge, G.E., J.S. Mansfield, and S.T. Chung. "Psychophysics of reading. XX. Linking letter recognition to reading." *Vision Res.* 2001 Mar; 41(6):725-43.

Loftus, G.R. and D.E. Irwin. "On the relations among different measures of visible and informational persistence." *Cognit. Psychol.* 1998 Mar;35[2]:135-99.

Logthetis, N.K. "Functional imaging of the monkey brain." *Nat. Neurosci.* 1999 Jun; [6]:555-62. Mednick 2002, *Percept area and resolution of perceptual images adversely compared by Kosslyn 1995.*

Majaj, N.J., D.G. Pelli, P. Kurshan, and M. Palomares. "The role of spatial frequency channels in letter identification." *Vision Res.* 2002 Apr; 42(9):1165-84.

Makeig S., T.P. Jung, and T.J. Sejnowski. "Awareness during drowsiness: dynamics and electrophysiological correlates." *Can. J. Exp. Psychol.* 2000 Dec; 54(4):266-73.

Maquet, P. "Experience-dependent changes in cerebral activation during human REM sleep." *Nat. Neurosci.* 2000 Aug; 3(8):831-packages 6.

Maquet, P. "The role of sleep in learning and memory." *Science.* 2001 Nov 2; 294(5544):1048-52. Wellcome Department of Cognitive Neurology, University College London, London WC1N 3BG, UK.

Mast, F.W. and S.M. Kosslyn. "Eye movements during visual mental imagery." *Trends Cogn. Sci.* 2002 Jul 6(7):271-272. Dept. of Psychology, Harvard University, 02138, Cambridge, MA, USA.

Moutoussis, K. and S. Zeki. "The relationship between cortical activation and perception investigated with invisible stimuli." *Proc. Natl. Acad. Sci.* USA 2002 Jul 9; 99(14):9527-32.

Ojanpaa, H., R. Nasanen, and I. Kojo. I "# Eye movements in the visual search of word lists." *Vision Res.* 2002 un; 42(12):1499-512.

Pernet, C., J. Uusvuori, and R. Salmelin. 2007. "Parafoveal-on-foveal and foveal word priming are different processes: behavioral and neurophysiological evidence."

Petersen, S. "Activation of Extrastriate and Frontal Cortical areas by visual words and word-like stimuli." *Science.* Aug. 1990 p 1041-2.

Pinker, S. "Mental imagery and the third dimension." *J. Exp. Psychol. Gen.* 1980 Sep; 109(3):354-71.

Polk, T.A. "The neural development and organization of letter recognition from functional neuroimaging, computational modelling and behavioral studies." *Proc. Nat. Acad. Sci. USA* Feb 3; 95(3):847-52.

Poldrack, R.A., J. Clark, E.J. Pare-Blagoev, D. Shohamy, J. Creso Moyano, C. Myers, and M.A. Gluck. "Interactive memory systems in the human brain." *Nature*. 2001 Nov 29; 414(6863):546-50. *Subjects relied upon the medial temporal lobe early in learning. However, this dependence rapidly declined with training.*

Pylyshyn, Z. "The role of location indexes in spatial perception: a sketch of the FINST spatial-index model." *Cognition*. 1989 Jun; 32(1):65-97.

Roland, A. and B. Gulyas. "Visual memory, visual imagery, and visual recognition of large field patterns by the human brain: functional anatomy by positron emission tomography." *Cortex*. 1995 Jan-Feb; 5(1):79-93.

Ross, J., M.C. Morrone, M.E. Goldberg, and D.C. Burr. "Changes in visual perception at the time of saccades." *Trends Neurosci*. 2001 Feb; 24(2):113-21.

Rubin, G.S. "Reading without saccadic eye movements." *Vision Res*. 1992 May: 32(5): 895-902.

Sandberg, A., A. Lansner, K.M. Petersson, and O. Ekeberg. "A Bayesian attractor network with incremental learning." *Network*. 2002 May; 13(2):179-94. New information to overwrite old, as in a *so-called palimpsest memory*.

Salvemini, A.V., A.L. Stewart, and D.G. Purcell. "The effects of foveal load and visual context on peripheral letter recognition."

Acta. Psychol. (Amst). 1996 Aug; 92(3):309-21. *Science.* 2001 Nov 2; 294(5544):1052-7.

Sejnowski, T.J. "Why do we sleep?" *Brain Res.* 2000 Dec 15; 866(1-2):208-223.

Siegel, J.M., P.R. Manger, R. Nienhuis, H.M. Fahringer, and J.D. Pettigrew. "Monotremes and the evolution of rapid eye movement sleep." *Philos. Trans. R. Soc. Lond. B. Biol. Sci.* 1998 Jul 9; 353(1372):1147-57.

Simmers, A.J. and P.J. Bex. "Deficit of visual contour integration in dyslexia." *Invest. Ophthalmol. Vis. Sci.* 2001 Oct; 42(11):2737-42. Distracter elements. Department of Visual Rehabilitation, Institute of Ophthalmology, University College London, United Kingdom. a.simmers@ucl.ac.uk.

Sinha, P. "Recognizing complex patterns." *Nat. Neurosci.* 2002 Nov; 5 Suppl:1093-7.

"Current artificial systems do not match the robustness and versatility of their biological counterparts."

Stickgold, R., J.A. Hobson, R. Fosse, and M. Fosse. "Sleep, learning, and dreams: off-line memory reprocessing." *Science.* 2001 Nov 2; 294(5544):1052-7.

Spielman, A.J. "Intracerebral hemodynamics between wakefulness and sleep." *Brain Res.* 2000 Jun 2; 866(1-2):313-25.

Steriade, M., I. Timofeev, and F. Grenier. "Natural waking and sleep states: a view from inside neocortical neurons." *J. Neurophysiol.* 2001 May; 85(5):1969-85.

Tassi, P. and A. Muzet. "Defining the States of Consciousness." *Neurosci. Bio. Behavior. Rev.* 2001: Mar; 25(2); 175-91.

Tulvig, E. and D. Schachter. "Priming and Human Memory Systems." 1990. *Science.* 247; 301-306.

Turrigiano, G.G. "Hebb and homeostasis in neural plasticity." *Curr. Opin. Neurobiol.* 2000 Jun; 10(3):358-64.

Tyler, W.J., M. Alonso, C.R. Bramham, and L.D. Pozzo-Miller. "From acquisition to consolidation: on the role of brain-derived neurotrophic factor signaling in hippocampal-dependent learning." *Learn Mem.* 2002 Sep; 9(5):224-37.

Uusitalo, M.A. "Dynamical organization of the human visual system revealed by lifetimes of activation traces." 1996. *Neuroscience Letters.* 213 149-152.

Van Hulle, M.M. "Kernel-based topographic map formation achieved with an information-theoretic approach." *Neural Netw.* 2002 Oct-Nov; 15(8-9):1029-39.

Wolfson, S.S. and Graham. "Comparing increment and decrement probes in the probed-sinewave paradigm." *Vision Res.* 2001 Apr; 41(9):1119-.

Whitaker. "Detection and discrimination of curvature in foveal and peripheral vision." *Vision Res.* 1993 Nov; 33(16):2215-24.

Xing, J. and D.J. Heeger. "Center-surround interactions in foveal and peripheral vision." *Vision Res.* 2000; 40(22):3065-72.

Topic C6 - Circus Ternus

Alais, D and J. Lorenceau J. “Perceptual grouping in the Ternus display: evidence for an ‘association field in apparent motion.” *Vision Res.* 2002 Apr; 42(8):1005-16.

Unite de Neurosciences Integratives et Computationnelles, UPR 2191 CNRS, Avenue de la Terrasse, Gif-sur-Yvette 91198, France Bex, P.J., A.B. Metha, and W. Makous. “Enhanced motion aftereffect for complex motions.” *Vision Res.* 1999 Jun; 39(13):2229-38. *That MAEs for radiation and rotation were greater than those for translation.*

Center for Visual Science, University of Rochester, NY 14627-0268, USA. pbex@essex.ac.uk Breitmeyer, Bruno G. and Alysia Ritter. 1986. “Visual Persistence and the Effect of Excentric Viewing, Element Size, and Frame Duration on Bi-stable Stroboscopic motion patients.” *Perception and Psychophysics.* 39 (4) 275-80.

Brown. “Spatial scale and cellular substrate of contrast adaptation by *retinal* ganglion cells.” *Nat. Neurosci.* 2001 Jan;4(1): 44-51. Found that “*The time required for contrast-adaptation varied with stimulus size, ranging from approximately 100 ms for the smallest stimuli, to seconds for stimuli the size of the receptive field.*”

Coombes, S. and P.C. Bressloff. “Mode locking and Arnold tongues in integrate-and-fire neural oscillators.” Q Delta Nonlinear and Complex Systems Group, Department of Mathematical Sciences, Loughborough University, Leicestershire LE11 3TU, United Kingdom.

Francis, G. and H. Kim. "Perceived motion in orientational afterimages: direction and speed."

Vision Res. 2001 Jan 15;41(2):161-72 Purdue University, Department of Psychological Sciences, 1364 Psychological Sciences Building, West Lafayette, IN 47907-1364, USA.

Gorea, A. and T.E. Conway. "Interocular interactions reveal the *opponent structure of motion* mechanisms." *Vision Res.* 2001 Feb;41(4):441-8 gfrancis@psych.purdue.edu.

Hock, H.S. and G.W. Balz.1994. "Organized motion patterns." *Vision Res.* 1994 Jul;34(14):1843-61. Spatial scale dependent in-phase and anti-phase directional biases in the perception of self-motion.

Hock, H.S., C.L. Park, and G. Schoner. "Self-organized pattern formation: experimental dissection of motion detection and motion integration by variation of attentional spread." *Vision Res.* 2002 Apr;42(8):991-1003. Department of Psychology, Florida Atlantic University, 33431, Boca Raton, FL, USA.

Hock, H.S., J.A. Kelso, and G. Schoner G. "Bistability and hysteresis in the organization of apparent *motion* patterns."

He, Z.J. and T.L. Ooi. "Perceptual organization of apparent motion in the Ternus display. Perception." 1999;28(7):877-92. Department of Psychology, University of Louisville, KY 40292, USA.

Kramer, Peter and Steven Yantis. 1997. "Perceptual Grouping in Space and Time: Evidence from the Ternus Display." *Perception and Psychophysics.* 59(1) 87-99.

Meese, T.S. and S.J. Anderson. "Spiral mechanisms are required to account for summation of complex motion components." *Vision Res.* 2002 Apr;42(9):1073-80. *With "both cardinal and spiral mechanisms with a half-bandwidth of approximately 46 degrees."*

Neurosciences Research Institute, School of Life and Health Sciences, Aston University, B4 7ET, Birmingham, UK Muller and Krauskopf; 2001, "Information conveyed by onset transients in responses of striate cortical neurons." *J. Neurosci.* 2001. Sep 1;21(17):6978-90. Which states, "In most neurons the onset of a grating gives rise to a transient discharge that decays with a time constant of 100 msec or less."

Ramachandran, V.S. and S.M. Anstis. "Perceptual organization in multistable apparent motion." *Perception.* 1985;14(2):135-43.

van de Grind's 2001 in "Slow and fast visual motion channels..." Comparisons between retinal and cortical receptive field-functions can be made.

Snowden, R.J. and F.A. Verstraten. "Motion transparency: making models of motion perception." *Trends Cogn. Sci.* 1999 Oct;3(10):369-377.

Stark, J. Neural Networks, *Learning Automata and Iterated Function Systems*, in: Fractals and Chaos, Springer-Verlag, New York, 1990. Persistence and the Effect of Eccentric Viewing, Element Size, and Frame Duration on Bistable Stroboscopic motion patients. Perception and Psychophysics. 9 (4) 275-280.

Tiesinga, P.H. "Precision and reliability of periodically and quasiperiodically driven integrate-and-fire neurons." *Phys. Rev.*

E. Sta.t Nonlin. Soft Matter Phys. 2002 Apr; Sloan-Swartz Center for Theoretical Neurobiology and Computational Neurobiology Laboratory, Salk Institute, 10010 North Torrey Pines Road, La Jolla, California 92037, USA.

Tso, D.Y. and C.D. Gilbert. "The organization of chromatic and spatial interactions in the primate striate cortex." 1: *J. Neurosci.* 1988 May;8(5):1712-27. That "blobs "contain red/green sensitive neurones, and the inter-blobs, the yellow/blue sensitive.

Bressloff, P.C., 280.

Tyler, C.W., H. Chan, and L. Liu. "Different spatial tunings for ON and OFF pathway stimulation." *Ophthalmic Physiol. Opt.* 1992 Apr;12(2):233-.

Topic C7 - Dynamic Motion-Memory Engrams

Baloch and Grossberg. "A neural model of high-level motion processing: line motion and form-motion dynamics." *Vision Res.* 1997 Nov;37(21):3037-59.

Bartels, Zeki S. "The theory of multistage integration in the visual brain." 1998 Dec. Proc. R. Soc. Lond. Biol. Soc. *Suggesting that "staged processing "(eg . of motion) is perceptually explicit*

Beard, B.L., S.A. Klein, and T. Carney. "Motion thresholds can be predicted from contrast discrimination." *J. Opt. Soc. Am. A.* 1997 Sep;14(9):2449-70.

Bex, P.J., A.B. Metha, and W. Makous. "Psychophysical evidence for a functional hierarchy of motion processing mechanisms." *J. Opt. Soc. Am. A. Opt. Image Sci. Vis.* 1998 Apr;15(4):769-76.

Bex, P.J. "Enhanced motion after-effect for complex motions." *Vision Res.* 1999 Jun;39(13) :2229-38.

Blake, R., N.J. Cepeda, and E. Hiris. "Memory for visual motion. *(30 sec).*" *Exp. Psychol.*

Hum. Percept Perform 1997 Apr;23(2):353-69.

Bouman, M.A. and W.A. van de Grind. 2000. "Motion coherence detection as a function of luminance in human central vision." *Vision Research.* 40, 3599-3611]. Result from recruitment of different sets of motion detectors, and an adjustment of their temporal properties.

Bowling. "The effects of spatial frequency and contrast on visual persistence." *Perception* 1979 8(5)529-39 *Low contrast gratings more persistent than high.*

Burr, D.C. and L. Santoro. "Temporal integration of optic flow, measured by contrast and coherence thresholds." *Vision Res.* 2001 Jul;41(15):1891-9.

Brady, N., P.J. Bex, and R.E. Fredericksen. "Independent coding across spatial scales in moving fractal images." *Vision Res.* 1997 Jul;37(14):1873-83.

Braundt, T. 1999. "The vestibular cortex, its location, functions and disorders." 1: Ann NY Acad. Sci. 1999 May 28;927293-312.

Brecht, M. 2001. "Coherent motion stimuli, responses in…supr. colliculus."

Rowe, J. "The prefrontal cortex response selection or maintenance within working memory." June 2000. 288. *Science.*

Castet, E. "The inverse- intensity effect is not lost with stimuli in apparent motion." *Vision Research* 1993 Aug. 33(12):169-708.

Clifford. 1999. "The perception and discrimination of speed in complex motion." *Vision Res.* 1999 Jun;39(13):2213-27. "*...speed discrimination appears to be based upon the pooled responses of elementary motion detectors...*"

Cornelissen, P.L. "Coherent motion detection and letter position encoding." *Vision Res.* 1998 Jul;38(14):2181-91. *Magnocellular pathway.*

Dixon, P. and V. Di Lollo. "Beyond visible persistence: an alternative account of temporal integration and segregation in visual processing." *Vision Res.* 2001;41(25-26):3505-11. *The correlation in time between the visual responses to the leading and trailing displays.*

Ffytche, D.H., C.N. Guy, and S. Zeki S. "The parallel visual motion inputs into areas V1 and V5 of human visual cortex." *Brain* 1995 Dec: 118(Pt6) 1375-94.

Francis, G. and H. Kim. "Perceived motion in orientational after- images: direction and speed." *Vision Res.* 2001 Jan 15;41 (2): 161-72.

Gegenfurtner, K.R. "Motion perception at scotopic light levels." *J. Opt. Am. A.* 2000 Sept;17(9): 1505-15.

Grossman, E.D. and R. Blake. "Brain Areas Active during Visual Perception of Biological Motion." *Neuron* 2002 Sep 12;35(6):1167-75.

Grunewald, A. "Orthogonal motion after-effect illusion predicted by a model of cortical motion processing." 1: *Nature* 1996 Nov 28:384(6607):358-60.

Guy, C.N. and S. Zeki. "The parallel visual motion inputs into areas V1 and V5 of human visual cortex." *Brain* 1995 Dec: 118(Pt6) 1375-94. *States that with stimulus speed moving at 22 degrees per sec signals reach V5, by-passing V1...but below 6 degrees per sec, the signals first reach V1.*

Hartley, A.A. and N.K. Speer. "Locating and fractionating working memory using functional neuroimaging: storage, maintenance, and executive functions." *Microsc. Res. Tech.* 2000 Oct 1;51(1):45-53.

Hoyer, P.O. and A. Hyvarinen. "A multi-layer sparse coding network learns contour coding from natural images."

Hess, R.F., P.J. Bex, E.R. Fredericksen, and N. Brady. "Is human motion detection subserved by a single or multiple channel mechanism?" *Vision Res.* 1998 Jan;38(2):259-66.

He, Z.J. "Perceptual organization of apparent motion in the Ternus display." *Perception* 1999;28(7):877-92.

Irwin, D.E. "Duration of visible persistence in relation to stimulus complexity." *Percept Psychophys.* 1991 Nov;50(5): 475-89.

Ishihara, M. "Effect of luminance contrast on the motion after-effect." *Percept. Mt. Skills* 1999 Feb;88(1):215-23. *Low contrast grating (5 percent) most effective @ 0.8 cycle degree.*

Joibes, Irwin, and Yantis. "Integrating visual information from successive fixations." 1982 *Science* Jan 8; 2135(4529):192-24 4.

Kramer, P. "Visible persistence and form correspondence in Ternus apparent motion." *Percept Psychophys.* 1999 Jul;61(5): 952-62.

Lankheet, M.J. "Spatio-temporal tuning of motion coherence detection at different luminance levels." *Vision Research* 2002 Jan;42(1):65-73.

Lepage, M., A.R. McIntosh, and E. Tulving. "Transperceptual encoding and retrieval processes in memory: a PET study of visual and haptic objects."

Mather, G. and S. Anstis. "Second-order texture contrast resolves ambiguous apparent motion." *Perception* 1995;24(12): 1373-82.

Mutoussis, K. and S. Zeki. "A direct demonstration of perceptual asynchrony in vision." *Proc. R. Soc. Lond. B. Biol. Sci.* 1997 Mar 22;264(13809):393-9. *The* perception *of color precedes that of motion perception "by some 80 ms."*

Poldrack, R.A. and M.G. Packard. "Competition among multiple memory systems: converging evidence from animal and human brain studies." *Neuropsychologia* 2003;41(3):245-51. Department of Psychology and Brain Research Institute, Franz Hall, University of California at Los Angeles, P.O. Box 951563, 90095-1563, Los Angeles, CA, USA.

Seiffert. "Position-displacement, not velocity is the cue to motion detection of *second* order stimuli." *Vision Res.* 1998 Nov;38 (22):3569-82.

Scott-Samuel, N.E. and R.F. Hess. "What does the Ternus display tell us about motion processing in human vision?" *Perception* 2001;30(10):1179-88.

Snowden, R.J. "Phantom -motion after-effects of detectors for the analysis of optic flow." *Curr. Biol.* 1997 Oct 1:7(10):717-22.

Snowden, R.J. "Visual attention to color: parvocellular guidance of attentional resources?" *Psychol. Sci.* 2002 Mar;13(2):180-4.

Shorter, S. and R. Patterson. "The stereoscopic (cyclopean) motion after-effect is dependent upon the temporal frequency of adapting mechanisms." *Vision Res.* 2001 June;41(14):1609-16. Smith, A.T. and T. Ledgeway. "Motion detection in human vision: a unifying approach based on energy and features." *Proc. R. Soc. Lond. BIOL. Sci.* 2001 Sept 22;268(1479): 1899-99.

Sperling and Lu. "Three-systems theory of human visual motion perception: review and update." Sept 2001 *J. Opt. Soc. Am.* 2331-70.

Tulving, E. and D. Schacter. "Priming and Human Memory Systems." *Science* 1990 Jan 19 1990 301-6 "Priming?"

Tulving in *Neuroimage* 2001 Sept; 14(3):527-84 *discusses one form of priming as a preperceptual representational system.*

Trezona, P. "The After-effects of a White Light Stimulus." *J. Physiol.* (1960) 154. Pp 67-78.

Ukkonen, O.I. and A.M. Derrington. "Motion of contrast-modulated gratings is analysed by different mechanisms at low and at high contrasts." *Vision Res.* 2000;40(24):3359-71.

Van de Grind, W.A., P. van Hof, M.J. van der Smagt, and F.A. Verstraten. 2001 "Slow and fast visual motion channels have independent binocular-rivalry stages." *Proc. R. Soc. Lond. B. Biol. Sci.* 2001 Feb 22;268(1465):437-43.

Verstraten, F.A., R. E. Fredericksen, R.J. Van Wezel, M.J. Lankheet, and W.A. Van de Grind. "Recovery from adaptation for dynamic and static motion aftereffects: evidence for two mechanisms." *Vision Res.* 1996 Feb;36(3):421-4.

Vidnyanszky, Z., E. Blaser, and T.V. Papathomas. "Motion integration during motion aftereffects." *Trends Cogn. Sci.* 2002 Apr 1;6(4):157-161.

Watamaniuk, S.N. "The human visual system averages speed information." *Vision Res.* 1992 May;32(5):931-41.

Zanker, J.M. "Perceptual deformation induced by visual." *Naturwissensenschaften* 2001 Mar;88(3):129-32.

Zeki, Fong et al. and D.J. McKeefry et al. "The activity in human areas VI/V2, V3 and V5 during the perception of coherent and incoherent motion." *Neuroimage.* Jan 1997; 5(1).

Braundt, T. "The vestibular cortex, its location, functions and disorders." 1: Ann NY Acad. Sci. 1999 May 28;927293-312 This describes intimate interaction with visual cortex to the prevailing mode of stimulation, body acceleration, (vestibular input) versus constant velocity motion (visual input).

Lewald, J., W.H. Ehrenstein, and R. Guski. "Spatio-temporal constraints for auditory-visual integration." *Behav. Brain Res.* 2001 Jun;121(1-2):69-79. Suggests bimodal neurons such as have been demonstrated by neurophysiological recordings in midbrain and cortex.

Salinas, E. and T.J. Sejnowski. "Gain modulation in the central nervous system: where behavior, neurophysiology, and computation meet." *Neuroscientist* 2001 Oct;7(5):430-40. One input, the modulatory one, affects the gain or the sensitivity of the neuron to the other input, without modifying its selectivity or receptive field properties. Department of Neurobiology and Anatomy, Wake Forest University School of Medicine, Winston-Salem, NC 27157, USA. esalinas@wfubmc.edu

Shams, L., Y. Kamitani, and S. Shimojo. "Visual illusion induced by sound." *Brain Res. Cogn. Brain. Res.* 2002 Jun;14(1):147-52. The temporal window of these audio-visual interactions is approximately 100 ms.

Watanabe, K. and S. Shimojo. "When sound affects vision: effects of auditory grouping on visual motion perception." *Psychol. Sci.* 2001 Mar;12(2):109-16. Is a genuine cross-modal effect.

Bowling. "The effects of spatial frequency and contrast on visual persistence." *Perception* 1979 ;8(5)529-39. (Low contrast gratings more persistent than high at 12 cycles per degree gratings, persistence **de**creased as contrast **in**creased.)

Ffytche, D.H. and M. Ishihara. "Effect of luminance contrast on the motion after-effect." *Percept. Mt. Skills.* 1999

Feb;88(1):215-23. *Low contrast grating (5 percent) was most effective @ 0.8 cycle degree.*

Castet, E. "The inverse-intensity effect is not lost with stimuli in apparent motion." *Vision Research* 1993 Aug. 33(12):169-708.

Lankheet, M.J. "Spatio-temporal tuning of motion coherence detection at different luminance levels." *Vision Res.* 2002 Jan;42(1):65-73.

Topic D12 - Rotation and Spiral Motions

Loose, R. and T. Probst. "Angular Velocity, not acceleration of self motion, mediates vestibular-visual interaction." *Perception* 2001;30(4)511-18 (4):719-25.

Morone, M.C. "Cardinal directions for visual optic flow." *Curr. Biol.* 1999 July15;9(14):763-6

"These are best tuned to "radial and rotational motion..."

Meese, T.S. "Broad direction bandwidths for complex motion mechanisms." *Vision Res.* 2001 Jul;41(15):1901-14. Favors expansions and rotations Poldrack, R.A. and M.G. Packard. "Competition among multiple memory systems: converging evidence from animal and human brain studies." *Neuropsychologia* 2003;41(3):245-51.

Shorter, S. and R. Patterson. "The stereoscopic (cyclopean) motion after-effect is dependent upon the temporal frequency of adapting mechanisms." *Vision Res.* 2001 June ; 41 (14):1609-16. Smith, A.T. and T. Ledgeway. "Motion detection in human vision: a unifying approach based on energy and features." *Proc. R. Soc. Lond. Biol. Sci.* 2001 Sept 22;268(1479): 1899-99.

Mather, G. and S. Anstis. "Second-order texture contrast resolves ambiguous apparent motion." *Perception* 1995;24(12):1373-82.

Zeki 1998, 1999, and 2001, "The autonomy of the visual systems and the modularity of conscious vision."

Mutoussis, K. and S. Zeki. "A direct demonstration of perceptual asynchrony in vision." *Proc. R. Soc. Lond. B. Biol. Sci.* 1997 Mar 22;264(13809):393-9. The *perception* of color precedes that of motion perception "by some 80 ms." (In our IMPPI phenomena color is not evident.)

Vidnyanszky, Z., E. Blaser, and T.V. Papathomas. "Motion integration during motion aftereffects." *Trends Cogn. Sci.* 2002 Apr 1;6(4):157-161. "The perceived global motion direction during motion aftereffects results from local vector averaging of the co-localized motion-direction signals induced by adaptation."

Salinas, E. and T.J. Sejnowski. "Gain modulation in the central nervous system: where behavior, neurophysiology, and computation meet." *Neuroscientist* 2001 Oct;7(5):430-40 "Gain modulation is revealed when one input, the modulatory one, affects the gain or the sensitivity of the neuron to the other input, without modifying its selectivity or receptive field properties."

Department of Neurobiology and Anatomy, Wake Forest University School of Medicine, Winston-Salem, NC 27157, USA. esalinas@wfubmc.edu.

Topic D9 - Migraine

Brown, in 2001, "Spatial scale and cellular substrate of contrast adaptation by retinal ganglion cells." *Nat. Neurosci.* 2001 Jan;4(1): 44-51 states: "The time required for contrast adaptation varied with stimulus size, ranging from approximately 100 ms for the smallest stimuli, to seconds for stimuli the size of the receptive field."

Bressloff 2002 to anisotropic long- range lateral connections via the LGN) (The third dimension of this latticed array is believed subjectively detectable..)

Hill, 1989, stated that 95 percent of the pulsatile component from ocular blood flow is in the choroid.

David Mackay (1977) refers to "Spatial Visual Noise."

Ivankov, N.Y. andn S.P. Kuznetsov. "Complex periodic orbits, renormalization, and scaling for quasiperiodic golden-mean transition to chaos." *Phys. Rev. E. Stat. Phys. Plasmas Fluids Relat. Interdiscip. Topics* 2001 Apr;63(4 Pt 2):046210. Demonstration of self-similarity on two-dimensional diagrams of Arnold tongues requires the use of a properly chosen curvilinear coordinate system. Saratov State University, Astrakhanskaja 83, Saratov, 410026, Russian Federation.

Manahilov, V., W.A. Simpson, and McCulloch. "Differences in pattern detectability across space in the central and peripheral visual fields. Spatial summation of peripheral Gabor patches." *J. Opt. Soc. Am. A. Opt. Image Sci. Vis.* 2001 Feb;18(2):273-82.

Sinderman, F., in 1969, briefly depicts the "Subjective development of the Eigengrau during and after dark adaptation to diffuse light excitation." 1: *Pflugers Arch.* 1969 ;307(2):142.

Tyler, in 1992, suggested that "stimuli without abrupt luminance transients reveal pronounced differences in the spatial tuning of responses to positive and negative stimuli, which probably reflect differences in the neural connectivity of the ON and OFF processing systems."

Bonhoeffer, T. and A. Grinvald. "The lay-out of iso-orientation domains in area 18 of cat visual cortex optical imaging reveals a pin-wheel-like organization." *J. Neurosci.* 1993 Oct; 13(10) :4157-80.

Crotogino, J. "Perceived scintillation rate of migraine aura." *Headache* 2001 Jan 1;41(1):40-48.

Gebhardt, M. "A cellular automaton model of excitable media (demonstrating)...including curvature and dispersion." *Science* Mar 30 1990 1563-6.

Grusser, O.J. "Migraine phosphenes and the retino-cortical magnification factor." 1: *Vision Res.* 1995 April;35(8):1125-34.

Hadjikhani et al. "Mechanisms of migraine aura revealed by fMRI in human visual cortex." April 10, 2001.

Knill, D.C. "Surface orientation from texture: ideal observers, generic observers and the information content of texture cues." *Vision Res.* 1998 Jun; 38 911: 1655-82.

Lewis, T.J. "Self-organized synchronous oscillations in a network of excitable cells coupled by gap junctions." *Network* 2000 Nov;11(4):299-320.

MacKay, D. "Interaction of stabilized retinal patterns with spatial visual noise." *Nature* pp.715-8.

Miller, E.K. *Neuron* 22,15 1999.

Reggia, J.A. and D. Montgomery. "1 A computational model of visual hallucinations in migraine." *Comput. Biol. Med.* 1996 Mar;26(2):133-41.

Swindale, N.V. "Cortical organization; modules, poly-maps and mosaics." 1: *Curr. Biol.* 1998 April 9;8 98): R270-3.

Swindale, N.V. *Cereb. Cortex* 2000 July;10(7):633-43. Questioned "How many maps are there in the visual cortex?"

Shmuel, A. and A. Grunvald. "Co-existence of linear zones and pin-wheels within orientational maps in cat visual cortex." *Pro. Natl. Sci. USA* 2000 May 9;97(10) 5568-73.

Worgotter, E. "A parallel noise-robust algorithm to recover depth information from radial flow fields." *Neural. Comput.* 1999 Feb 15;11(2):391-416.

Rubin, G.S. and K. Turano. "Reading without saccadic eye movements." *Vision Res.* 1992 May; 32(5) 895-902.

Jonides, J. *Science* 1982 Jan 8;215[4529]:192-4.

Topic D10 - Mandala

Bressloff, P.C. and J.D. Cowan. "An amplitude equation approach to contextual effects in visual cortex." *Neural. Comput.* 2002 Mar;14(3):493-525.

Brown, C. and J. Gebhard. "Visual field articulation in the absence of spatial stimulus gradients." Nat. Neurosci. 2001 Jan;4(1):44-51. "Spatial scale and cellular substrate of contrast adaptation by retinal ganglion cells."

Brown, C. "Spatial scale and cellular substrate of contrast adaptation by retinal ganglion cells." *Nat. Neurosci.* 2001 Jan ;4(1): 44-51. *States: "The time required for contrast adaptation varied with stimulus size, ranging from approximately 100 ms for the smallest stimuli, to seconds for stimuli the size of the receptive field."*

Curcio, C.A. "Packing geometry of human cone photoreceptors: variation with eccentricity and evidence of anisotropy." *Vis. Neurosci.* 1992 Aug. 9(2):169-80. Elliot C. @. U. Washington Oct 25-7 2001 on functional image synthesis. Swirling per C.E. "The further the point from the origin the more the motion."

Enright, J.T. "On the "Cyclopean eye and the reliability of perceived straight ahead." *Vision Res.* 1998 Feb: 38(3):459-69.

Heeley, D.W. and B. Timney. "Human photoreceptor topography." *J. Comp. Neurol.* 1990 Feb. 22:2 (4) 497-523. Describes a streak of cone high-density along one rung, the horizontal meridian; this microscopic study supported subjectively by 1989 Vision Research. Hill, 1989, stated that 95 percent of the pulsatile component from ocular blood flow in the choroid.

Ivankov, N.Y. and S.P. Kuznetsov. "Complex periodic orbits, renormalization, and scaling for quasiperiodic golden-mean transition to chaos." *Phys. Rev. E. Stat. Phys. Plasmas Fluids Relat. Interdiscip. Topics* 2001 Apr;63(4 Pt 2):046210. *Demonstration of self-similarity on two-dimensional diagrams of Arnold tongues requires the use of a properly chosen curvilinear coordinate system.* Saratov State University, Astrakhanskaja 83, Saratov, 410026, Russian Federation.

Khan, A.Z. and J.D. Crawford. 2001.

Knill, D.C. "Surface orientation from texture: ideal observers, generic observers and the information content of texture cues." *Vision Res.* 1998 Jun; 38 911: 1655-82-416.

Li, A. and Q. Zaidi. *Vision Res.* 2000;40(2):217-42. With this conclusion: "Perception of three-dimensional shape from texture is based on patterns of oriented energy."

Mackay (1977) referred to it as "Spatial Visual Noise."

Manahilov, V., W.A. Simpson, and McCulloch. "Differences in pattern detectability across space in the central and peripheral visual fields/ Spatial summation of peripheral Gabor patches." J. *Opt. Soc. Am. A. Opt. Image Sci. Vis.* 2001 Feb;18(2):273-82.

Mumford, D. "Discriminating figure from ground : the role of edge detection and region growing." 1987 *Proc. Natl. Acad. Sci. USA* Oct ;84(20):7354-8. *With reference to Eigengrau perceptions, it seems intuitively that the luminances constitute the figure.* F. Sinderman, in 1969, alludes to the "Subjective development of the Eigengrau during and after dark adaptation to diffuse light excitation." 1: *Pflugers Arch.* 1969 ;307(2):142.

Tyler, C.W., H. Chan, and L. Liu. "Different spatial tunings for ON and OFF pathway stimulation." *Ophthalmic Physiol. Opt.* 1992 Apr;12(2):233-40. *"Stimuli without abrupt luminance transients reveal pronounced differences in the spatial tuning of responses to positive and negative stimuli, which probably reflect differences in the neural connectivity of the ON and OFF processing systems."*

Wilson, H.R. "A neural model of foveal light adaptation and after-image formation." *Vis. Neurosci.* 1997 May-Jun; 14(3):403-23. *The M and P ganglion cells pathways and also both the ON- and OFF-centers contribute, it is said, to a negative after-image perception.*

Worgotter, E. "A parallel noise-robust algorithm to recover depth information from radial flow." *Neural. Comput.* 1999 Feb 15;11(2):391-.

Topic D11 - Helmholtz Traveling Waves (HTW)

Akopian, A. "Neuromodulation of ligand- and voltage-gated channels in the amphibian retina." *Microsc. Res. Tech.* 2000 Sep 1;50(5):403-10. *A high degree of adaptability, retinal synapses have evolved multiple neuromodulatory mechanisms.*

Burioka, N. "Relationship between correlation dimensions and indices of liner analysis in both respiratory movement and electroencephalogram." *Clin. Neurophysiol.* 2001.

Bazhenov, M., I. Timofeev, M. Steriade, and T.J. Sejnowski. "Model of thalamocortical slow-wave sleep oscillations and transitions to activated states." *J. Neurosci.* 2002 Oct 1;22(19):8691-704. Coupled map lattices.

The Salk Institute, Howard Hughes Medical Institute, Computational Neurobiology Laboratory, La Jolla, California 92037, USA. bazhenov@salk.edul;112(7) :1147-53o?? Dacey DM "Horizontal cells of the primate retina: *cone specificity* without *spectral opponency.*" Dacey, D.M. *Science* 1996 Dec 20 Feb 2 ;274(5295):656-9 and 1999Feb 2 ;2271(5249):616-7 *Are we seeing differential firing sequences arising from these two populations of horizontal cells?* Rudolph, M., J.M. Fellous, and T.J. Sejnowski. "Fluctuating synaptic conductances recreate in vivo-like activity in neocortical neurons." *Neuroscience* 2001;107(1):13-24. *Represent the currents generated by thousands of stochastically releasing synapses. The presence of high-amplitude membrane potential fluctuations, a low-input.*

Unite de Neurosciences Integratives et Computationnelles, CNRS, UPR-2191, Gif-sur-Yvette, France. destexhe@iaf.cnrs-gif.fr Helmholtz H Handbook of Physiological Optics Optical Soc Amer 1924 Li A Zaidi Q. Vision Res. 2000;40(2):217-42., , . *"Perception of three-dimensional shape from texture is based on patterns of oriented energy* "

Hock, H.S., J.A. Kelso, and G. Schoner. "Bistability and hysteresis in the organization of apparent *motion* patterns." *J. Exp. Psychol. Hum. Percept. Perform.* 1993 Feb;19(1):63-80.

In relation to illustory arrays/images?

Department of Psychology, Florida Atlantic University, Boca Raton 33431. Hoyuelos, M., D. Walgraef, P. Colet, andn M. San Miguel. "Patterns arising from the interaction between scalar and vectorial instabilities in two-photon resonant Kerr cavities." *Phys. Rev. E. Stat. Nonlin. Soft Matter Phys.* 2002 Apr;65(4 Pt 2B):046620. Linearly polarized hexagonal pattern whereas the other instability is of pure vectorial origin and would give rise to an elliptically polarized stripe pattern.

Marshal, L. "Slow potential shifts at sleep-wake transitions between NREM and REM sleep." *Sleep* 1996 Feb. 19(2): 145-51. *Scalp recordings in transitional states show negative DC shifts lasting 190 seconds, suggesting a temporary phase of increased cortical excitability.*

Saunders, R.D. and J.G. Jefferys. "Weak electric field interactions in the central nervous system." *Health Phys.* 2002 Sep;83(3):366-75.

Timofeev, I., F. Grenier, M. Bazhenov, T.J. Sejnowski, and M. Steriade. "Origin of slow cortical oscillations in deafferented cortical slabs." *Cereb. Cortex* 2000 Dec;10(12):1185-99. *The isolated slabs displayed brief active periods separated by long periods of silence, up to 60 s in duration. larger territories, both the frequency and regularity of the slow oscillation approached that generated in intact cortex*s.

Tso, D.Y. and C.D. Gilbert. "The organization of chromatic and spatial interactions in the primate striate cortex." 1: *J. Neurosci.* 1988 May;8(5):1712-27. *Contrary to previous reports, few true double color-opponent cells were found. Some blob cells* red/green color opponency were three *times more numerous than blue/yellow blobs.*

Laboratory of Neurobiology, Rockefeller University, New York, New York 10021.

Wilson, H.R., R. Blake, and S.H. Lee. "Dynamics of travelling waves in visual perception." *Nature* 2001 Aug 30;412(6850):907-10. *Termed binocular rivalry, these fluctuating states of perceptual dominance and suppression are thought to provide a window into the neural dynamics that underlie conscious visual awareness.*

Topic D12 – Rotating Spiral Waves

Alonso, S., F. Sagues, and J.M. Sancho. "Target patterns created out of noise and traveling or spiral waves sustained by noise." *Phys. Rev. E. Stat. Nonlin. Soft Matter Phys.* 2002 Jun;65(6 Pt 2):066107.

Agladze, K. "Light induced annihilation and shift of spiral waves." *Chaos.* 1996 Sep;6(3):328-333. Epub 2002 Jun 14.

B. Julesz, "Texture and Visual Perception," *Sci. Am.* 212, 38–48 (1965).

Bar, M., A.K. Bangia, and I.G. Kevrekidis. "Bifurcation and stability analysis of rotating chemical spirals in circular domains: Boundary-induced meandering and stabilization." *Phys. Rev. E. Stat. Nonlin. Soft Matter Phys.* 2003 May;67(5-2):056126. Epub 2003 May 27.

D. Mihalache, D. Mazilu, L.C. Crasovan, I. Towers, B.A. Malomed, A.V. Buryak, L. Torner, F. Lederer, "Stable three-dimensional spinning optical solitons supported by competing quadratic and cubic nonlinearities," *Phys. Rev. E. Stat. Nonlin. Soft Matter Phys.* 66 (2002).

D. Charalampidis, "Texture synthesis: textons revisited," *IEEE Trans. Image Process.* 15, 777–87 (2006).

Dahlem, M.A. and S.C. Muller. "Self-induced splitting of spiral-shaped spreading depression waves in chicken retina." *Exp. Brain Res.* 1997 Jun;115(2):319-24. Trajectory (extension approximately 1.2 mm) that is described by the tip over one spiral revolution (period 2.45+/-0.1 min).

Duerk, J.L. "The goal of any acquisition strategy is to map k-space completely, and two methods, spiral imaging and echo-planar imaging, are described to demonstrate that the data acquisition path need not be a straight line." *Magn. Reson. Imaging Clin. N. Am.* 1999 Nov;7(4):629-59.

Gottwald, G., A. Pumir, and V. Krinsky. "Spiral wave drift induced by stimulating wave trains." *Chaos* 2001 Sep;11(3):487-494. The drift of meandering spirals...The property of meandering of spirals is not robust against periodic stimulations.

Guo, H., H. Liao, and Q. Ouyang. 2002 "Relation between the wave front and the tip movement of spirals." "The measurement shows that the slope of the linear R-lambda curve is about 0.13, independent of the control parameter."

H.X. Hu, L. Ji, Q.S. Li, "Delay-induced inward and outward spiral waves in oscillatory medium," *J. Chem. Phys.* 128 (2008).

Henry, H. and V. Hakim. 2002. "Different bands of modes are seen to be unstable in different regions of parameter space. Spatial symmetry breaking in the Belousov-Zhabotinsky reaction with light-induced remote communication." *Phys. Rev. Lett.* 2001 Aug 20;87(8):088303. Epub 2001 Aug 06.

Hildebrand, M., H. Skodt, and K. Showalter. "Complex behavior of colliding and splitting wave fragments is found with feedback radii comparable to the spiral wavelength." *Phys. Rev. Lett.* 2001 Aug 20;87(8):088303. Epub 2001 Aug 06.

J.L. Cantero, M. Atienza, R. Stickgold, M.J. Kahana, J.R. Madsen, B. Kocsis, "Sleep-dependent theta oscillations in the human hippocampus and neocortex," *J. Neurosci.* 23, 10897–903 (2003).

Jensen, F.G., J. Sporring, M. Nielsen, and P.G. Sorensen. "Tracking target and spiral waves." *Chaos* 2002 Mar;12(1):16-26. "The center of the pattern...a new concept, called the spiral focus, which is defined by the evolutes of the actual spiral or target wave."

L. Condat, D. Van De Ville, B. Forster-Heinlein, "Reversible, fast, and high-quality grid conversions," *IEEE Trans. Image Process.* 17, 679–93 (2008).

Leisman, G., and P. Koch. "The neural continuum has the property of amplifying waves of wavelength large compared with synaptic connection ranges." *Int. J. Neurosci.* 2003 Feb;113(2):181-204.

M. Conrad, "The geometry of evolution," *Biosystems* 24, (1990).

Matthews, P.C. "Pattern formation on a sphere." *Phys. Rev. E. Stat. Nonlin. Soft Matter Phys.* 2003 Mar;67(3-2):036206.

Nagy-Ungvarai, Z., J. Ungvarai, and S.C. Muller. "Complexity in spiral wave dynamics(a)." *Chaos* 1993 Jan;3(1):15-19. Within a (concentration, time) parameter plane, the movement of free ends of waves was classified as follows: (a) in a stable domain-periodic rigid rotation with cores of small (200 &mgrm) or very large (2 mm) diameter; quasiperiodic compound motion along a hypocycle, a straight loopy line or an epicycle; complex meandering composed of possibly more than two components; (b) rectilinear tip motion indicating the boundary of spiral wave stability; and (c) in an unstable domain-shrinking of open ends of wave fronts during propagation.

Namba, T., T. Ashihara, K. Nakazawa, and T. Ohe. 1999 "A meandering spiral wave."

Nayagam, V.V. and F.A. Williams. "Rotating spiral edge flames in von karman swirling flows." *Phys. Rev. Lett.* 2000 Jan 17;84(3):479-82. These flames exhibit similarities to patterns commonly found in quiescent excitable media.

Ouyang, Q., H.L. Swinney, and G. Li. "Transition from spirals to defect-mediated turbulence driven by a doppler instability." *Phys. Rev. Lett.* 2000 Jan 31;84(5):1047-50.

S. Ishihara, K. Kaneko, "Turing pattern with proportion preservation," *J. Theor. Biol.* 238, 683–93 (2006).

Seipel, M., F.W. Schneider, and A.F. Munster. 2001. "Control and coupling of spiral waves in excitable media."

Steinbock, O. and S.C. Muller. "Spatial attractors in aggregation patterns of Dictyostelium discoideum." *Z. Naturforsch [C].* 1995 Mar-Apr;50(3-4):275-81. Curve (radius approximately 130 microns) corresponding to the boundary of the spiral core. A vortex-like rotation of cells is found close to the core of spiral waves, with a maximum velocity of 15 microns/min. Cell motion and spiral tip orbiting follow an opposite sense of rotation.

T. Heimburg, A.D. Jackson, "On soliton propagation in biomembranes and nerves," *Proc. Natl. Acad. Sci. U.S.A.* 102, 9790–5 (2005).

Takagi, S., A. Pumir, L. Kramer, and V. Krinsky. "Mechanism of standing wave patterns in cardiac muscle." *Phys. Rev. Lett.* 2003 Mar 28;90(12):124101. Epub 2003 Mar 26.

W. Park, G.S. Chirikjian, "Interconversion between truncated Cartesian and polar expansions of images," *IEEE Trans. Image Process.* 16, 1946–55 (2007).

Y.A. Astrov, "Phase transition in an ensemble of dissipative solitons of a Turing system,"

Phys. Rev. E. Stat. Nonlin. Soft Matter Phys. 67 (2003).

Part E13–16 - Bistable Hues

Bill, A. and G. Sperber. "Control of retinal and choroidal blood flow." *Eye* 1990 4 [Pt2]319-25 *choroidal blood vessels…where autoregulation is absent.*

Department of Physiology and Medical Biophysics, University of Uppsala, Sweden.

Earlier studies on the control of retinal and choroidal blood flow are reviewed and some recent observations on the effects of light on retinal metabolism and retinal and choroidal blood flow in monkeys (Macaca fascicularis) are reported in preliminary form. *The retina is nourished by the retinal blood vessels, where blood flow is autoregulated and the* choroidal blood vessels where autoregulation is absent. Studies with the deoxyglucose method of Sokoloff indicate that flickering light tends to increase the metabolism of the inner retina, while constant light reduces the metabolism in the outer retina. Retinal blood flow in flickering light, 8 Hz, is higher than in constant light.

The sympathetic nerves of the choroid are probably involved in a protective mechanism, preventing overperfusion in fight and flight situations with acute increments in blood pressure. The facial nerve contains parasympathetic vasodilator fibres to the choroid; the physiological significance of these fibres is unknown. The neuropeptides NPY, VIP and PHI are likely to be involved in Autonomic reflexes in the eye.

Vis. Research 18. 1623-31.

Brown, C. et al. "Entoptic perimetry screening for central diabetic scotomas and macular edema." *Ophthalmology* 2000 107;4, 755-759.

Brown, K.T. and J. Crawford. "Melanin and the rapid light evoked response from the pigment epithelium of the frog eye." *Vision Res.* 1967 7. 165-178 Pergamon Press.

Buerk, D.G., C.E. Riva, and S.D. Cranstoun. "Frequency and luminance-dependent blood flow and K ion exchanges during flicker stimuli in cat optic nerve head." *Invest.Ophthalmol. Vis. Sci.* 1995 36[11]:2216-27.

Buxton R, B. and L.R. Frank. "A model for the coupling between cerebral blood flow and oxygen metabolism during neural stimulation." *J. Cereb. Blood Flow Metab.* 1997 Jan;17(1):64-72. Disproportionately large changes in blood flow are required in order to support small changes in the O2 metabolic rate consistent with tight coupling of flow and oxidative metabolism.

Chapman, C.L., J.J. Wright, and P.D. Bourke. "Spatial eigenmodes and synchronous oscillation: co-incidence detection in simulated cerebral cortex." *J. Math Biol.* 2002 Jul;45(1):57-78.

"Cortical activation levels are increased, local damped oscillations in the gamma band undergo a transition to highly nonlinear undamped activity with 40 Hz dominant frequency."

Cristini, C., D. Forlani, and C. Scardovi. "Choroidal Circulation in Glaucoma." *Brit. J. Ophth.* 1962 46 99. *(40 percent decrease in choroidal thicckness by age 75.)*

Crittin, M., H. Schmidt, and C.E. Riva. "Hemoglobin oxygen saturation (So2) in the human ocular fundus measured by reflectance oximetry: preliminary data in retinal veins." *Klin Monatsbl Augenheilkd* 2002 Apr;219(4):289-91.

Delaey, C. and J. Van De Voorde. "Regulatory mechanisms in the retinal and choroidal circulation." *Ophthalmic Res.* 2000 Nov-Dec;32(6):249-56. *The choroidal circulation is mainly controlled by sympathetic innervation and is not autoregulated.*

Department of Physiology and Physiopathology, University of Gent, Belgium.

The retina receives its nutrients from two separate circulations: retinal and choroidal circulation. This short overview describes the determinants in the regulation of these circulations. Retinal circulation is characterized by a low blood flow while flow in the choroid is high. *The choroidal circulation is mainly controlled by sympathetic innervation and is not autoregulated.* Retinal circulation lacks autonomic innervation, shows an efficient autoregulation and is mainly influenced by local factors. Local mediators released by endothelial cells and surrounding retinal tissue also have a substantial role in the regulation of retinal circulation. Copyright 2000 S. Karger AG, Basel Duann, J.R., T.P. Jung, W.J. Kuo, T.C. Yeh, S. Makeig,

J.C. Hsieh, and T.J. Sejnowski. "Single-trial variability in event-related BOLD signals." *Neuroimage* 2002 Apr;15(4):823-35. Hemodynamic response (HR) to 8-Hz flickering-checkerboard stimulation were presented for 0.5-s (short) or 3-s (long) durations at 30-s intervals.

Computational Neurobiology Laboratory, The Salk Institute for Biological Studies, La Jolla, California 92037, USA. Eisner, A. "Samples R. Flicker Sensitivity and Cardiovascular Function in Healthy Middle-aged people." *Arch. Ophthalmol.* 2000: 118:1049-1055.

Falsini, B., C.E. Riva, and E. Logean.

"Flicker-evoked changes in human optic nerve blood flow: relationship with retinal neural activity." *Invest. Ophthalmol. Vis. Sci.* 2002 Jul;43(7):2309-16.

Fisahn, A., F.G. Pike, E.H. Buhl, and O. Paulsen. "Cholinergic induction of network oscillations at 40 Hz in the hippocampus in vitro." *Nature.* 1998 Jul 9;394(6689):132-3.

Fry, G.A. "The Bezold-Brucke phenomena at the two ends of the spectrum." *Am. J. Optom. Physiol. Opt.* 1983 Dec;60(12):977-81. *Study of the purple colors involving mixtures of 460 and 667 nm.*

Jandrasits, K., A. Luksch, G. Soregi, G.T. Dorner, K. Polak, and L. Schmetterer.

"Effect of noradrenaline on retinal blood flow in healthy subjects." *Ophthalmology* 2002 Feb;109(2):291-5. *Adrenergic system appears* not to *play a major role in retinal blood flow regulation.*

Gebhard, J. W. "Chromatic Phenomena produced by intermittent stimulation of the retina." *J. Exp. Nov 1947 Psych.* 33:387-4O06 p 402. "*...a ubiquitous violet cloud floating about.*" Grunwald, J.E., S.M. Hariprasad, J. DuPont, M.G. Maguire, S.L. Fine, A.J. Brucker, A.M. Maguire, and A.C. Ho. "Foveolar choroidal blood flow in age-related macular degeneration." *Invest. Ophthalmol. Vis. Sci.* 1998 Feb;39(2):385-90.

Harris, A., C. Jonescu-Cuypers, B. Martin, L. Kagemann, M. Zalish, and H.J. Garzozi. "90 Simultaneous management of blood flow and IOP in glaucoma." *Acta. Ophthalmol. Scand.* 2001 Aug;79(4):336-41. Further, *glaucoma patients with normal IOP show clear evidence for cerebral and ocular ischemia.*

Harris, A., L. Kageman, and G. Cioffi. "Assessment of Human Ocular Hemodynamics." *Survey of Ophthalmology* 198 509-533.

Herrmann, C.S. "Human EEG responses to 1-100 Hz flicker: resonance phenomena in visual cortex and their potential correlation to cognitive phenomena." *P. Exp. Brain Res.* 2001 Apr;137(3-4):346-53. Resonances at 10, 20, 40, 80 Hz. Max-Planck Institute of Cognitive Neuroscience, Postfach 500 355, 04303 Leipzig, Germany.

Jefferys, J.G. "Neuronal networks for induced 40 Hz. rhythms." *Trends Neurosci.* 1996 May;19(5) :202-8.

Krizaj, D. and P. Witovsky. "Effects of submicromolar concentrations of dopamine on photoreceptor to horizontal cell comunication." *Brain Res.* 1993Nov 5;627[1];122-3.

Polak, K., L. Schmetterer, and C.E. Riva. "Influence of flicker frequency on flicker-induced changes of retinal vessel diameter." *Invest. Ophthalmol. Vis. Sci.* 2002 Aug;43(8):2721-6

In retinal arteries an increase was observed at all flicker frequencies, with a less-pronounced effect at 64 Hz. In retinal veins, all flicker frequencies except 2 and 64 Hz induced vasodilation.

Polska, E., A. Luksch, P. Ehrlich, A. Sieder, and L. Schmetterer. "Measurements in the peripheral retina using LDF and laser interferometry are mainly influenced by the choroidal circulation." *Curr. Eye Res.* 2002 Apr;24(4):318-23. *Laser with either of the two methods is focused onto the fovea it is obvious that only choroidal blood flow contributes to the signals.*

Pournaras, C.J. and C.E. Riva CE. "Studies of the hemodynamics of the optic head nerve using laser Doppler flowmetry." *Ophtalmol.* 2001 Feb;24(2):199-205."*...demonstrated for the first time a dynamic coupling of blood flow to function and metabolism in the ONH, mediated by an increase in potassium and nitric oxide release.*"

Marshall, C.R. "Entoptic phenomena associated with the retina." *Brit. J. Ophth.*1935 19: 177-201.

Orge, F., A. Harris, L. Kagemann, K. Kopecky, C.W. Sheets, E. Rechtman, and M. Zalish. "The first technique for non-invasive measurements of volumetric ophthalmic artery blood flow in humans." *Br. J. Ophthalmol.* 2002 Nov;86(11):1216-9. There was no correlation with the central retinal artery. No, it's all local autonomy.

Polak, K., L. Schmetterer, A. Luksch, S. Gruber, E. Polska, V. Peternell, and Bayerle. "Free fatty acids/triglycerides increase ocular and subcutaneous blood flow." *Am. J. Physiol. Regul. Integr. Comp. Physiol.* 2001 Jan;280(1):R56-61. Plasma FFA/*triglyceride elevation induced a rise in pulsatile choroidal blood flow by 25 +/- 3 percent (P < 0.001) and in retinal blood flow by 60 +/- 23 percent* Department of Clinical Pharmacology, Department of Internal Medicine III, University of Vienna, A-1090 Vienna, Austria.

Rager, G. and W. Singer. "The response of cat visual cortex to flicker stimuli of variable frequency." *Eur. J. Neurosci.* 1998 May;10(5):1856-77. *The gamma-frequency range between 30 and 50 Hz. About 300 ms after flicker onset, responses stabilized and exhibited a highly regular oscillatory patterning that was surprisingly similar at different recording sites due to precise stimulus locking.*

University Fribourg, Institute of Anatomy, Fribourg, Switzerland. Rennie, C.J., J.J. Wright, and P.A. Robinson. "Mechanisms of cortical electrical activity and emergence of gamma rhythm." *J. Theor. Biol.* 2000 Jul 7;205(1):17-35. *A resonance in the gamma range near 40 Hz.*

Ross, R. et al. "Presumed Macular Choroidal Watershed." *A. J. O.* 1998125;71-80.

Speilman, A.J. et al. "Intracerebral hemodynamics probed by near infra-red spectroscopy in the transitions between wakefulness and sleep." *Brain Res.* 2000 Jun2;866[1-2]:313-25.

Madsen, P.L. et al. "Cerebral O2 metabolism and cerebral blood flow in humans during deep and rapid eye movement sleep." *J. Appl. Physiol.* 1991 Jun; 70[6]; 2597-601.

Shurcliff, W. "A new visual phenomenon..." *J. Optic Soc. Am.* 1959 49.11. 1041- 48.

Smith, V.C., J. Pokorny, B.B. Lee, and D.M. Dacey. "Primate horizontal cell dynamics: an analysis of sensitivity regulation in the outer retina." *J. Neurophysiol.* 2001 Feb;85(2):545-58. *A resonance shoulder near 40 Hz.*

Vo Van, Toi and C. Riva. "Variations of blood flow at optic nerve head induced by sinusoidal flicker ref..."

Welpe, E. "Der Violetteffect, eine Neue Durch Flimmerlicht Induzierte Subjective Farbe." 1978 *Vis. Research* 18. 1623-31.

Topic E14 - Ocular Blood Flow

Eisner, A. "Multiple components in photopic dark adaptation." *J. Opt. Soc. Am. A.* 1986 May;3[5]:655-66.

Wang, L., M. Kondo, and A. Bill. "Glucose metabolism in cat outer retina. Effects of light and hyperoxia." *Investig. Ophth. & Vis. Science* 1997 38[1];48-55.

Brown, C. et al. "Entoptic perimetry screening for central diabetic scotomas and macular edema." *Ophthalmology* 2000 107;4, 755-759.

Buerk, D.G., C.E. Riva, and S.D. Cranstoun. "Frequency and luminance-dependent blood flow and K ion exchanges during flicker stimuli in cat optic nerve head." *Invest.Ophthalmol. Vis. Sci.* 1995 36[11]:2216-27.

Speilman, A.J. et al. "Intracerebral hemodynamics probed by near infra-red spectroscopy in the transitions between wakefulness and sleep." *Brain Res.* 2000 Jun2;866[1-2]:313-25.

Madsen, P.L. et al. "Cerebral O2 metabolism and cerebral blood flow in humans during deep and rapid eye movement sleep." *J. Appl. Physiol.* 1991 Jun; 70[6]; 2597-601.

Hsu, S. and R. Molday. "Glucose metabolism in photoreceptor outer segments." *Journal of Biol. Chem.* 1994.269[27]:1795-59.

Buerk D.G. and C.E. Riva. Adenosine enhances functional activation of blood flow in cat optic nerve head during photic stimulation independently from nitric oxide.

Microvasc Res. 2002 Sep;64(2):254-64. PMID: 12204650 [PubMed - in process]

Polak, K, Schmetterer L, Riva CE. Influence of flicker frequency on flicker-induced changes of retinal vessel diameter. Invest Ophthalmol Vis Sci. 2002 Aug;43(8):2721-6.

PMID: 12147608 [PubMed - indexed for MEDLINE]

Falsini B, Riva CE, Logean E. Flicker-evoked changes in human optic nerve blood flow: relationship with retinal neural activity. Invest Ophthalmol Vis Sci. 2002 Jul;43(7):2309-16.

PMID: 12091432 [PubMed - indexed for MEDLINE]

Falsini B, Riva CE, Logean E. Relationship of blood flow changes of the human optic nerve with neural retinal activity: a new approach to the study of neuro-ophthalmic disorders.

Klin Monatsbl Augenheilkd. 2002 Apr;219(4):296-8.

PMID: 12022022 [PubMed - indexed for MEDLINE]

Crittin M, Schmidt H, Riva CE. Hemoglobin oxygen saturation (So2) in the human ocular fundus measured by reflectance oximetry: preliminary data in retinal veins. Klin Monatsbl Augenheilkd. 2002 Apr;219(4):289-91.

PMID: 12022020 [PubMed - indexed for MEDLINE]

Logean E, Falsini B, Riva CE. [Effect of chromatic flicker on circulation of the optic nerve]

Klin Monatsbl Augenheilkd. 2001 May;218(5):345-7. French.

PMID: 11417332 [PubMed - indexed for MEDLINE]

Riva CE, Falsini B, Logean E. Flicker-evoked responses of human optic nerve head blood flow: luminance versus chromatic modulation. Invest Ophthalmol Vis Sci. 2001 Mar;42(3):756-62.

PMID: 11222538 [PubMed - indexed for MEDLINE]

Geiser MH, Riva CE, Dorner GT, Diermann U, Luksch A, Schmetterer L. Response of choroidal blood flow in the foveal region to hyperoxia and hyperoxia-hypercapnia.

Curr Eye Res. 2000 Aug;21(2):669-76.

PMID: 11148604 [PubMed - indexed for MEDLINE]

Longo A, Geiser M, Riva CE. Subfoveal choroidal blood flow in response to light-dark exposure. Invest Ophthalmol Vis Sci. 2000 Aug;41(9):2678-83.

PMID: 10937582 [PubMed - indexed for MEDLINE]

Logean E, Schmetterer LF, Geiser MH, Riva CE. [Optical Doppler velocimetry of red blood cells at different depths in retinal vessels by varying the coherence length of the source: feasibility study] Klin Monatsbl Augenheilkd. 2000 May;216(5):313-5. French.

PMID: 10863702 [PubMed - indexed for MEDLINE]

Longo A, Geiser M, Riva CE. [Effect of light on choroidal blood flow in the fovea centralis]

Klin Monatsbl Augenheilkd. 2000 May;216(5):311-2. French.

PMID: 10863701 [PubMed - indexed for MEDLINE]

Riva CE, Logean E, Petrig BL, Falsini B. [Effect of dark adaptation on retinal blood flow]

Klin Monatsbl Augenheilkd. 2000 May;216(5):309-10. French.

PMID: 10863700 [PubMed - indexed for MEDLINE]

Topic E15 - The Prompt Ocular-Perfusion Marker Offers Applications in Glaucoma Management

Brown and Gebhard in 1947 "Visual field articulation in the absence of spatial stimulus gradients the inner machine"

Gorea, A., T.E. Conway, and R. Blake. "Interocular interactions reveal the opponent structure of motion mechanisms." *Vision Res.* 2001 Feb 41(4):441-8.

Hock, H.S., J.A. Kelso, and G. Schoner. "Bi-stability and hysteresis in the organization of apparent *motion* patterns." *J. Exp. Psychol. Hum. Percept. Perform.* 1993 Feb;19(1):63-80. Krapivsky, P.L. and S. Redner S. "The interplay between the integration and inhibition leads to a steady state." *Phys. Rev. E. Stat. Phys. Plasmas Fluids Relat. Interdiscip. Topics* 2001 Oct;64(4-1):041906.

Shurcliff, W. "A new visual phenomenon." *J. Optic Soc. Am.* 1959 49.11. 1041-48.

Smythies. "The Stroboscopic Patterns I The Dark Phase." *British Journal of Psychology.* 50:106-16, 1959.

Sterzer, P., M.O. Russ, C. Preibisch, and A. Kleinschmidt. "Neural correlates of spontaneous direction reversals in ambiguous apparent visual motion." *Neuroimage* 2002 Apr;15(4):908-16.

Department of Neurology, Johann Wolfgang Goethe-University, Theodor-Stern-Kai 7, D-60590 Frankfurt am Main, Germany.

Suzuki, S. and M.A. Peterson. "Multiplicative effects of intention on the perception of bi-stable apparent motion." Psychol. Sci. 2000 May;11(3):202-9.

Tso, D.Y. and C.D. Gilbert. "The organization of chromatic and spatial interactions in the primate striate cortex." 1: *J. Neurosci.* 1988 May;8(5):1712-27.

Wachtler, T., T.W. Lee, and T.J. Sejnowski. "Chromatic structure of natural scenes." *J. Opt. Soc. Am. A. Opt. Image Sci. Vis.* 2001 Jan;18(1):65-77.

Welpe, E. "Der Violetteffect, eine Neue Durch Flimmerlicht Induzierte Subjective Farbe." 1978 *Vis. Research* 18. 1623-31.

Wilson, H.R. "A neural model of foveal light adaptation and after-image formation." *Vis. Neurosci.* 1997 May-Jun ; 14(3):403-23. *The M and P ganglion cells pathways and also both the on- and off-centers contribute, it is said, to a negative after-image perception.*

Topic F17 - Visual Perceptions of Vestibular Signals

Bedell, H.E. "Perception of a clear and stable world with congenital nystagmus." *Optom. Vis. Sci.* 2000 Nov;77(11):573-81. (The extra-retinal signals which accompany involuntary eye movements employ the same mechanisms as exist in the normal individuals in applying…lesser ongoing corrections for stabiliziation.)

Bressloff, P.C. "BLOCH waves, periodic feature maps, and cortical pattern formation." *Phys. Rev. Lett.* 2002 Aug 19;89(8):088101.

Department of Mathematics, University of Utah, Salt Lake City, Utah 84112. "...crystallinelike structure that breaks Euclidean symmetry to the discrete symmetry of a planar lattice group."

Burkell, J.A. and Z.W. Pylyshyn. "Searching through subsets: a test of the visual indexing hypothesis." *Spat. Vis.* 1997;11(2):225-58.

Burr, D. 2000. "Motion vision: are 'speed lines' used in human visual motion?"

Currie, C.B. 2000. "The role of the saccade target object in the perception of visually stable world."

De Waele, C. "Vestibular projections in the human cortex." *Exp. Brain Res.* 2001 Dec;141(4) 541.

Fendrich, R. and P.M. Corballis. "The temporal cross-capture of audition and vision." *Percept. Psychophys.* 2001 May;63(4):719-25. "Temporal discrepancies in the input from different sensory modalities are reconciled and could provide a probe for examining the neural stages at which evoked responses reach consciousness."

Galati, G., G. Committeri, J.N. Sanes, and L. Pizzamiglio. "Spatial coding of visual and somatic sensory information in body-centred coordinates." *Eur. J. Neurosci.* 2001 Aug;14(4):737-46.

Jell. "The influence of active versus passive head oscillation, and mental set on the human vestibulo-ocular reflex." *Environ. Med.* 1988 Nov; 59(11 pt 1):1061-5. "Using an imagined head-fixed target in the dark, VOR gain was near zero at low frequencies."

Laskhminarrayan, V. "A generalized perceptual space." 1: *Neurpol. Res.* 2000 Oct 22(7) 699-702.

Lu, Z. and G. Sperling. 'Three-systems theory of human visual motion perception: review and update." Sept 2001 *J. Opt. Soc. Am.* 2331-70. On "pedestals."

Mon-Williams, M., and J.R. Tressilian. "A framework for considering the role of afference and efference in the control and perception of ocular position." *Biol. Cybern.* 1999 Aug; 799(2):175-89.

Roth, J.D. and S.M. Kosslyn. "Construction of the third dimension in mental imagery." "*Cognit. Psychology* 1988 Jul;20(3):344-61.

Scholl, B.J. and K. Nakayama. "Causal capture: contextual of collision events." *Psychol. Sci.* 2002 Nov;13(6):493-8. Causal capture to occur, the context event need be present for only 50 ms surrounding the impact, "but capture is destroyed by only 200 ms of temporal asynchrony between the two events."

Tyler, C.W. and L.L. Kontsevich. "Stereoprocessing of cyclopean depth images: horizontally elongated summation fields." Vision Res. 2001 Aug;41(17):2235-43.

Wade, N.J. and M.T. Swanson. "A general model for the perception of space and motion." 1: *Perception* 1996 ;25 (2):187-94.

Van de Grind, W. "Physical, neural, and mental timing." *Conscious Cogn.* 2002 Jun;11(2):241. Vestibulo-ocular reflex function as measured with the head autorotation test. Hirvonen TP, Pyykko I, Aalto H, Juhola M.Department of Otorhinolaryngology, University Central Hospital of Helsinki, Finland.

Loose, R. and T. Probst. "Angular Velocity, not acceleration of self motion, mediates vestibular-visual interaction." *Perception* 2001;30(4)511-18 4):719-25.

Poldrack, R.A. and M.G. Packard. "Competition among multiple memory systems: converging evidence from animal and human brain studies." *Neuropsychologia* 2003;41(3):245-51.

Pylyshyn, Z.W. "Situating vision in the world." *Trends Cogn. Sci.* 2000 May;4(5):197-207.

Visual indexes, preconceptual objects, and situated vision.

Pylyshyn, Z.W. "Early vision must pick out and compute the relation among several individual objects while ignoring their properties…incrementally computing and updating representations of a dynamic scene."

Burkell, J.A. and Z.W. Pylyshyn. "Searching through subsets: a test of the visual indexing hypothesis." *Spat. Vis.* 1997;11(2):225-58.

Rutgers Center for Cognitive Science, Rutgers University, Psychology Building, New Wing, Busch Campus, New Brunswick, NJ 08903, USA. zenon@ruccs.rutgers.edu Tyler,

C.W. and L.L. Kontsevich. "Stereoprocessing of cyclopean depth images: horizontally elongated summation fields." *Vision Res.* 2001 Aug;41(17):2235-43.

Topic F18 – Ambiguity and Biases

Albert, M.K. "Cue interactions, border ownership and illusory contours." *Vision Res.* 2001 Oct;41(22):2827-34. Perception of one or another illusory surface depends on the outcome of such a competition, the alternative percepts primarily exhibit bistability rather than averaging (or mutual weakening).

Department of Psychology, University of Southampton, Highfield, Southhampton SO17 1BJ, UK. mka@soton.ac.uk

Bartolomeo, P. "The relationship between visual perception and visual mental imagery: a reappraisal of the neuropsychological evidence." *Cortex* 2002 Jun;38(3):357-78. "...recent reports of patients showing double dissociations between perception and imagery abilities challenged the perception-imagery equivalence"

Bunge, S.A., E. Hazeltine, M.D. Scanlon, A.C. Rosen, and J.D. Gabrieli. "Dis-sociable contributions of prefrontal and parietal cortices to response selection." *Neuroimage* 2002 Nov;17(3):1562-71. Neurosciences Program, Stanford University, USA. These findings support the idea that parietal cortex is nvolved in activating possible responses on the basis of learned stimulus-response associations, and that prefrontal cortex is recruited when there is a need to select between competing responses.

Burr, D.C. and J. Ross. "Direct evidence that "speedlines "influence motion mechanisms." *J. Neurosci.* 2002 Oct 1;22(19):8661-4.

Cave, K.R., S. Pinker, L. Giorgi, C.E. Thomas, L.M. Heller, J.M. Wolfe, and H. Lin. "The representation of location in visual images." *Cognit. Psychol.* 1994 Feb;26(1):1-32. Vanderbilt University, Department of Psychology, Nashville, TN 37240.

Finke, R.A. and H.S. Kurtzman. "Mapping the visual field in mental imagery." *J. Exp. Psychol. Gen.* 1981 Dec;110(4):501-17. *That imagery and perceptual fields are very similar in shape, both being elongated horizontally and extending farther below the point to which one's gaze is directed than above.*

Mast, F.W. and S.M. Kosslyn. "Visual mental images can be ambiguous: insights from individual differences in spatial transformation abilities." *Cognition* 2002 Nov;86(1):57-70.

Mellet, E., N. Tzourio-Mazoyer, S. Bricogne, B. Mazoyer, S.M. Kosslyn, and M. Denis.

"Functional anatomy of high-resolution visual mental imagery." *J. Cogn. Neurosci.* 2000 Jan;12(1):98-109. GIN GIP Cyceron, BP 5229, 14074 Caen Cedex France. mellet@cyceron.fr

Pinker, S. "The representation of location in visual images." *Cognit. Psychol.* 1994 Feb;26(1):1-32. "*Images can display metric 2-D distance information in a perspective view never actually experiences, so mental images not simply be 'snapshot plus scale model' pairs. The three-dimensional structure of objects is encoded in long-term memory in 3-D object-centered coordinate systems.*"

Poldrack, R.A. and M.G. Packard. "Competition among multiple memory systems: converging evidence from animal and human brain studies." *Neuropsychologia* 2003;41(3):245-51.

Rentschler, I., M. Juttner, A. Unzicker, and T. Landis. "Innate and learned components of human visual preference." 1999, Institute of Medical Psychology, University of Munich, Goethestrasse 31, 80336, Munchen, Germany. ingo@imp.med.uni-muenchen.de. Aesthetic sense reflects preferences for image signals whose characteristics best fit innate brain mechanisms of visual recognition.

Sterzer, P., M.O. Russ, C. Preibisch, and A. Kleinschmidt. "Neural correlates of spontaneous direction reversals in ambiguous apparent visual motion." *Neuroimage* 2002 Apr;15(4):908-16.

Department of Neurology, Johann Wolfgang Goethe-University, Theodor-Stern-Kai 7, D-60590 Frankfurt am Main, German

Pernet, C., J. Uusvuori, and R. Salmelin. "Parafoveal-on-foveal and foveal word priming are different processes: behavioral and neurophysiological evidence." *Neuroimage* 2007.

Brain Research Unit, Low Temperature Laboratory, Helsinki University of Technology, Finland.

Parafoveal-on-foveal priming refers to the presentation of an item (the prime) in parafoveal vision followed by the presentation of an item (the target) in foveal vision In natural reading, the 'parafoveal preview benefit' subserves fluent reading as, e.g., reading times increase when such information is not available.

Yet, the neural correlates of reading are mostly studied with foveally presented stimuli and little is known of this parafoveal influence. Here, we used complementary information from a behavioral study and a magneto-encephalography experiment to clarify the relationship between parafoveal-on-foveal and foveal priming.

Unlike foveal priming, parafoveal-on-foveal priming was present only at short prime-to-target delay (<100 ms). Behaviorally, the parafoveal priming effect was influenced by the prime visual field (left/right) and target lexical type (word/nonword), suggesting emphasis on perceptual analysis for LVF primes and on conceptual analysis for RVF primes. At the neural level, the overall sequence of activation was similar for foveal and parafoveal primes followed by foveal word targets, but the priming effects were bilateral for foveal primes versus left-lateralized for RVF primes. No neural effects of priming appeared for LVF primes, in line with the RVF preference imposed by the Western writing system. These results highlight the role of the left hemisphere in linguistic analysis and point out possible limitations of foveal stimulus presentation for drawing conclusions about natural reading.

About the Author

Ida Pearce, M.D. received her medical degree at the Royal College of Surgeons of England, with a Diplomate in Ophthalmology in 1951. She completed clinical studies at Moorfields and the Royal Eye Hospital in London.

During the 1970s, she was involved with glaucoma research at UCLA and the Veterans Administration Hospital in Long Beach, California, but the majority of her career has been spent as a practicing ophthalmologist, with a private practice in Long Beach, California, and at Kaiser Permanente Medical Group, Department of Ophthalmology in Bellflower, California where she was a partner for over 30 years.

In addition to her interest in research and as a practicing clinician, Dr. Pearce is the president and founder of M.A.W. Inc., a nonprofit child care center, which she started over forty years ago with funding from St. Mary's and Memorial Hospitals in Los Angeles County to provide day care for nurses with infants who worked long shifts with odd hours.

Dr. Pearce started and ran the first nonsmoking clinic in the 1970s at Kaiser Permanente in Bellflower, California, and produced stress-released videotapes, which were played in the waiting rooms and hospital patients' rooms at Kaiser Permanente. She has often been described as a pioneer and ahead of her time. She has raised five children and has recently relocated to Orange County, California, where, at age ninety, she continues her interest in research and findings related to Memory Traces: Recursive Engrams.

Dr. Pearce can be contacted at ipearce@visualmemoryresearch.org.

Acknowledgments

I have been enlighted and encouraged by the writings of James McGaugh, Daniel Schacter, John Smythes, Brian Wandell, Richard Semon, and Leon Glass, to name a few authors. The physicians and staff at The Kaiser Permanente Medical Centers, notably Walter Schwimmer, M.D., Basilio Kalpakian, M.D., Jan Chapman, and Nancy Lusa, were supportive of my various projects including a non-smoking clinic and stress-relief videotapes.

I appreciated the help of my administrative colleagues including Katherine Carr, Jacqueline Carr, Judith Smith, Shramana Ghosh, MaryAnne Wendt, Leslie Uchino, and Roxan Shoa. Finally, I am grateful for the support of my family and friends during these endeavors.

Made in the USA
Middletown, DE
25 March 2023